国家"双高"建设立体化教材
全国船舶工业职业教育教学指导委员会"十三五"规划教材

船舶工程技术专业英语

主　编　杨文林
副主编　黄晓雪
主　审　刘　歆

哈尔滨工程大学出版社
Harbin Engineering University Press

内容简介

《船舶工程技术专业英语》一书是渤海船舶职业学院国家"双高"建设教材。全书共设置六个模块，分别是船舶设计、船舶性能、船体结构、船舶图纸、船舶建造、船舶检验与维护。教材对每个模块又进行分解，从而使学生可以学习到船舶从设计到生产全过程中所涉及的专业英语知识。每节课的内容都包括对学生听、说、读、写能力的锻炼，培养学生具有利用所学知识进行实际语言交流的能力。本书作为《船舶日常实用英语》的后续教材，可进一步强化学生的专业英语知识。

本书适用于高职类涉船专业学生，也可供船厂技术人员查阅。

图书在版编目（CIP）数据

船舶工程技术专业英语 / 杨文林主编． -- 哈尔滨：哈尔滨工程大学出版社，2022.4（2025.2重印）
ISBN 978-7-5661-3394-6

Ⅰ．①船… Ⅱ．①杨… Ⅲ．①船舶工程－英语－高等职业教育－教材 Ⅳ．①U66

中国版本图书馆CIP数据核字（2022）第043379号

船舶工程技术专业英语
CHUANBO GONGCHENG JISHU ZHUANYE YINGYU

选题策划	张志雯　薛　力
责任编辑	张　彦　王　宇
封面设计	李海波

出版发行	哈尔滨工程大学出版社
社　　址	哈尔滨市南岗区南通大街145号
邮政编码	150001
发行电话	0451-82519328
传　　真	0451-82519699
经　　销	新华书店
印　　刷	哈尔滨午阳印刷有限公司
开　　本	787 mm×1 092 mm　1/16
印　　张	8.5
字　　数	250千字
版　　次	2022年4月第1版
印　　次	2025年2月第3次印刷
定　　价	32.00元

http://www.hrbeupress.com
E-mail:heupress@hrbeu.edu.cn

前言 Foreword

随着船舶行业在我国的快速发展，行业对船舶专业人才的需求也逐步增多，对于船舶专业人才的要求也日趋国际化。依据渤海船舶职业学院"双高"建设项目方案总体目标，结合船舶工程技术专业群人才培养方案修订工作，共建共享型专业基础课和专业技能课程，进行《船舶工程技术专业英语》教材建设。

本书作为船舶工程技术专业的核心课程，在整个船舶行业人才的培养体系中占有非常重要的位置，为提升教材的实用性，本书遵循"立足岗位，实用为主，够用为度"的原则，结合船舶行业高职学生学习英语的特点和规律，力求选择难度适度、实用性强的文章，帮助学生在打好英语语言基本功的基础上，全面提升船舶行业英语的应用能力。

本书依据职业标准进行课程设置，首先分析船舶工程技术专业对应的岗位，根据岗位要求细化确定相应岗位所需具备的专业能力，进而确定六个教学模块，分别是船舶设计、船舶性能、船体结构、船舶图纸、船舶建造、船舶检验与维护模块。对每个模块又进行了分解，从而学生可以学习到船舶从设计到生产的全过程中所涉及的专业英语知识。每节课的内容中都包括对学生听说读写能力的锻炼，培养学生利用所学船舶工程技术专业英语进行实际语言交流的能力。根据岗位职责和工作流程，本书将岗位工作能力分解为15个单元，让学生在清楚理解船舶建造的各个工作岗位标准程序的基础上，能够逐个环节地掌握船舶专业英语表达技巧和岗位操作技能。课程总学时为30学时，每个单元2学时。

本书由渤海船舶职业学院杨文林、黄晓雪编写，由渤海造船厂集团公司船舶设计院刘歆高级工程师主审。全书由渤海船舶职业学院杨文林担任主编并编写了模块一至模块三以及词汇表的内容，黄晓雪担任副主编并编写了模块四至模块六的内容。在本书的编写过程中还得到了渤海船舶职业学院王建红、王璐璐、齐蕴思、王璞和船厂技术人员等同志的大力支持和帮助，在此一并致以衷心的感谢。

由于编者水平有限和时间仓促，疏漏之处在所难免，恳请广大同人批评斧正。

编　者
2022年3月

目录 contents

Module 1　Ship Design ·· 1
　Lesson 1　The Ship's Functions, Features and Types ···························· 1
Module 2　Ship Performance ··· 9
　Lesson 2　Principal Dimensions ·· 9
　Lesson 3　Sea Keeping Performance ·· 17
Module 3　Ship Structure ··· 26
　Lesson 4　Equipment to Ensure Stability ··· 26
　Lesson 5　Ship Structure (1) ··· 34
　Lesson 6　Ship Structure (2) ··· 43
Module 4　Ship Drawings ··· 52
　Lesson 7　Deck and Bulkhead Construction ··· 52
　Lesson 8　Lines Plan ·· 61
　Lesson 9　General Arrangement Plan ·· 70
　Lesson 10　Basic Structural Plan ·· 79
Module 5　Ship Production ·· 88
　Lesson 11　Ship Construction Process ·· 88
　Lesson 12　Fabrication and Welding of Hull (1) ··· 96
　Lesson 13　Fabrication and Welding of Hull (2) ··· 104
Module 6　Ship Survey and Quality Management ································· 112
　Lesson 14　Survey ··· 112
　Lesson 15　Quality Management ·· 120

Module 1 Ship Design

Lesson 1 The Ship's Functions, Features and Types

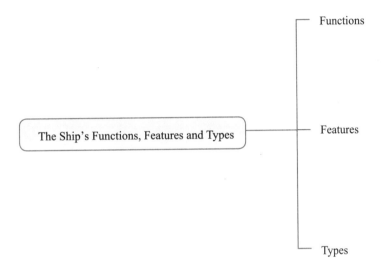

Background

A: Do you know how many types of ships can be divided?

B: According to different classification methods, ships can be classified into many types, and they can be simply classified into military ships and civilian ships. Next we will study in further detail.

Text reading

1. The Functions, Features of the Ships

Ship is a floating structure of water engineering. It is the main tool of water transportation, water work and water combat (作战). Ships play an important role in national defense (国防), national economy and marine development.

When a ship is navigating in water, the water condition is complicated. The hull must not only bear the weight of goods, machinery (机器) and equipment, the pressure of water, the impact of wind and waves and other external forces, but also possess reliable water-tightness and sufficient firmness. At the same time, it should have good performance, beautiful appearance and reasonable economic benefits.

In the course of ship development, there have been several significant changes in the materials and connecting methods of the components used in the hull structure. The first boats were canoe structure, which later evolved into a combination of planks and beams. In the 18th century, with the development of metallurgical industry(冶金工业) and mechanical manufacturing(机械制造业), ships with iron and iron-wood composite structure began to appear. In the second half of the 19th century, the use of low-carbon steel for shipbuilding increased, steel ships gradually replaced wooden ships and wood-iron mixed ships, and steel became the main material for shipbuilding. In recent decades, with the increase of ship size, high-strength steel is used in shipbuilding, which reduces the size of structural members and reduces the weight of the structure.

Since the 1930s, welding shipbuilding has replaced riveting. Welding makes the hull structure more complete, more compact and lighter than riveting. At present, steel boats are built by welding.

2. The Types of Ships

There are many ways to classify ships. According to thier sailing conditions, ships can be divided into sea-going ship(沿海，近海，远洋船), harbor boat(港湾船) and inland vessel(内河船); according to the status of navigation ships can be classified into displacement ship(排水型船), submarine, planing ship(滑行艇), hydrofoil boat(水翼艇), aerofoil boat(冲翼艇) and hovercraft; according to the way of navigation, they can be divided into selfpropelled vessel(自航船) and non-propelled ship; according to the number of hull, they can be divided into single-hull ships and multi-hull ship(多体船), catamaran(双体船) are more common in multi-hull ships.

The classification of ships by purpose can be broadly classified as follows:

（1）Transport ship—including general cargo ship(Fig.1.1), passenger ship, intermediate ship(客货船), ferry(渡船), container ship, roll on/roll off ship, barge carrier, barge, refrigerated ship, bulk cargo carrier, tank ship, chemical carrier, liquid gas carrier.

Fig. 1.1 General Cargo Ship

Lesson 1　The Ship's Functions, Features and Types　/3

（2）Engineering ship—including dredger（挖泥船）(Fig.1.2), floating crane（起重船）, cable ship（布缆船）, salvage ship（救捞船）, icebreaker（破冰船）, pile driving barge（打桩船）, floating dock, ocean development ship（海洋开发船）, drilling ship（钻井船）, drilling platform（钻井平台）and so on.

Fig. 1.2　Dredger

（3）Fishing ship—including drift fishing boat（拖渔网船）(Fig.1.3), angling boat（钓鱼船）, fishing guidance ship, fishing survey ship, fishing processing ship, whaling ship（捕鲸船）and so on.

Fig. 1.3　Drift Fishing Boat

（4）Habror boat—including tugboat (Fig.1.4), pilot boat（引航船）, fire boat（消防船）, depot ship（供应船）, accommodation ship（交通船）, etc.

Fig. 1.4　Tugboat

（5）Marine surveying ship—including marine synthetic surveying ship（Fig.1.5）, marine professional (hydrology, geology, biology) surveying ship, deep sea vehicle and so on.

Fig. 1.5　Marine Synthetic Surveying Ship

（6）Combat ship—including aircraft carrier（Fig.1.6）, cruiser, destroyer, frigate（护卫舰）, mine layer（布雷舰）, minesweeper（扫雷舰艇）, landing ship, submarine, submarine chaser（猎潜艇）and all kinds of speedboats.

Fig. 1.6　Aircraft Carrier

（7）Auxiliary ship—including replenishment oiler（补给舰）（Fig.1.7）, repair ship, training ship, degaussing vessel（消磁船）, hospital ship, etc.

Fig. 1.7　Replenishment Oiler

New Words and Expressions

1. transportation [ˌtrænspɔːrˈteɪʃn] *n.* 运送，运输；运输系统；运输工具
2. development [dɪˈveləpmənt] *n.* 发展，进化，新产品；开发区
3. navigate [ˈnævɪɡeɪt] *vi.* 航行；驾驶；操纵 *vt.* 航行；驾驶；横渡；
4. water-tightness 水密性
5. firmness [ˈfɜːmnəs] *n.* 坚固；坚牢；坚定
6. economic benefit 经济效益

7. significant [sɪgˈnɪfɪkənt] *adj.* 重大的；有效的；有意义的；值得注意的；意味深长的

8. evolve [ɪˈvɒlv] *vt.* 发展，进化；进化；使逐步形成；推断出

9. composite [ˈkɒmpəzɪt] *adj.* 合成的；复合的 *n.* 合成物；复合材料

10. low-carbon steel 低碳钢

11. riveting [ˈrɪvɪtɪŋ] *n.* 铆接（法）*adj.* 饶有趣味的，非常精彩的，引人入胜的

12. submarine [ˌsʌbməˈriːn] *n.* 潜水艇；海底生物 *adj.* 海底的；水下的

13. hovercraft [ˈhʌvərkræft] *n.* 气垫船

14. transport ship 运输船

15. engineering ship 工程船

16. fishing ship 渔业船

17. habror boat 港务船

18. marine surveying ship 海洋调查船

19. combat ship 战斗舰艇

20. auxiliary ship 辅助船舰

Notes

1. Ship is a floating structure of water engineering. It is the main tool of water transportation, water work and water combat（作战）.Ships play an important role in national defense（国防）, national economy and marine development.

   译：船舶是一种浮动的水上工程建筑物，是人们从事水上交通运输、水工作业和水中作战的主要工具。船舶在国防、国民经济和海洋开发等方面都占有十分重要的地位。

2. When a ship is navigating in water, the water condition is complicated. The hull must not only bear the weight of goods, machinery（机器）and equipment, the pressure of water, the impact of wind and waves and other external forces, but also possess reliable water-tightness and sufficient firmness.

   译：船舶在水中航行，水况复杂，船体不仅要承受货物与机器及设备的重力、水的压力、风浪的冲击等外力作用，还应具备可靠的水密性和足够的坚固性。

3. In the course of ship development, there have been several significant changes in the materials and connecting methods of the components used in the hull structure.

   译：在船舶发展过程中，船体结构在其所用的材料、构件的连接方法等方面，曾有几个重大的变革。

Expanding reading

History of Modern Shipping

The period from 1800 until the second world war saw the rise of the regular service liner. This was the result of the transport of cargo and passenger between Europe and the colonies in the East and the West, and the increasing number of emigrants leaving for North America.

Shipbuilding changed slowly but steadily to facilitate the new demands using new technologies.

The main developments were:

（1）Wood as main construction material was replaced by iron and later by steel.

（2）Sailing ships were replaced by steam ships and later by motor ships.

（3）New types of ships like tankers and reefers were developed.

In general, the large and versatile trading vessels of this period remained in use until the 1970s. Transportation of passengers, general cargo, oil, refrigerated cargo, heavy boxed parcels, animal and bulk with one and the same ship was very common. Even today's "multipurpose" ships do not achieve this level of versatility.

After some initial hesitation, the period after The World War Ⅱ showed a contumacious increase in global commerce, only interrupted by short periods of relapse, lasts even to this day. In the beginning this resulted in more and more ships, subsequently they became faster and bigger. A lot of smaller ships were then taken out of service. The modernization of shipbuilding and navigation led to the loss of many jobs in the sector. After the 1970`s more and more universal ships were replaced by specialized vessels that can carry only one type of cargo. This process had already started on a much smaller scale since 1900.

Passenger liners have been superseded almost entirely by aeroplanes, because of the large distances involved. However, after 1990 the number of passenger ships that specialize in luxury cruises has increased enormously.

New Words and Expressions

1. colony ['kɒləni] *n.* 殖民地
2. emigrant ['emɪgrənt] *n.* 移居外国的人，移民
3. facilitate[fəˈsɪlɪteɪt] *v.* 促进，促使，使便利
4. refrigerated cargo 冷藏货物
5. multipurpose[ˌmʌltiˈpɜ:pəs] *adj.* 多种用途的，多目标的
6. hesitation [ˌhezɪˈteɪʃn] *n.* 犹豫，疑虑，不情愿
7. contumacious[ˌkɑ:ntuˈmeɪʃəs] *adj.* 违抗的，不服从的
8. enormously[ɪˈnɔ:rməsli] *adv.* 非常，极其

Exercises

I. Answer the following questions according to the passage.

1. What types of ships can be classified according to their sailing conditions?

Lesson 1 The Ship's Functions, Features and Types

2. What types of ships can be classified according to the status of navigation ?
   _____

Ⅱ. **Practic these new words.**

1. English to Chinese.

   sea-going ship _____   harbor boat _____

   general cargo ship _____   aircraft carrier _____

   submarine _____   hospital ship _____

2. Chinese to English.

   气垫船 _____   运输船 _____

   工程船 _____   渔业船 _____

   港务船 _____   战斗舰艇 _____

Ⅲ. **Explain the nouns.**

1. ship

   _____

2. sea-going ship

   _____

Ⅳ. **Fill in the blanks with the proper words or expressions given below.**

> economic benefits, navigating, complicated, but also, bear

When a ship is _____ in water, the water condition is _____. The hull must not only _____ the weight of goods, machinery and equipment, the pressure of water, the impact of wind and waves and other external forces, _____ possess reliable water-tightness and sufficient firmness. At the same time, it should have good performance, beautiful appearance and reasonable _____.

Ⅴ. Translation.

1. Translate the following sentences into Chinese.

（1）When a ship is navigating in water, the water condition is complicated.

（2）At the same time, it should have good performance, beautiful appearance and reasonable economic benefits.

（3）In the course of ship development, there have been several significant changes in the materials and connecting methods of the components used in the hull structure.

（4）In recent decades, with the increase of ship size, high-strength steel is used in shipbuilding, which reduces the size of structural members and reduces the weight of the structure.

（5）Welding makes the hull structure more complete, more compact and lighter than riveting. At present, steel boats are built by welding.

2. Translate the short passage.

There are many ways to classify ships. According to their sailing conditions, ships can be divided into sea-going ship（沿海，近海，远洋船）, harbor boat and inland vessel; according to the status of navigation ships can be classified into displacement ship, submarines, planing ship, hydrofoil boat, aerofoil boat and hovercraft; according to the way of navigation, they can be divided into selfpropelled ship and non-propelled ship; according to the number of hull, they can be divided into single-hull ship and multi-hull ship, catamarans are more common in multi-hull ships.

Module 2　Ship Performance

Lesson 2　Principal Dimensions

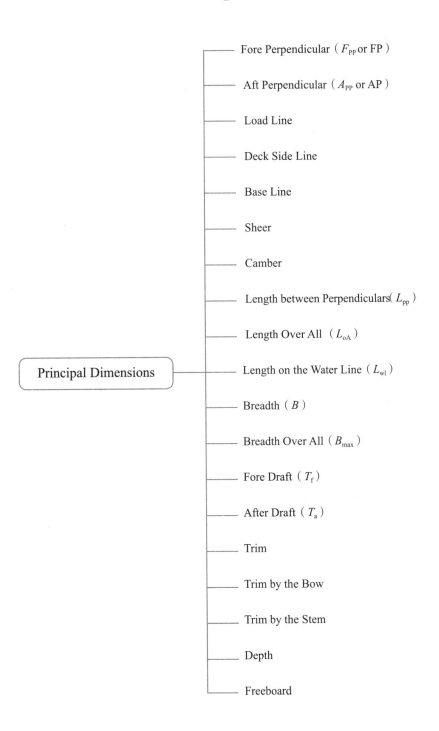

Background

A: Do you know what the principal dimensions of a ship include?

B: It includes length between perpendiculars, length over all, length on the water line, breadth and so on. Next we will study in further detail.

Text reading

All sea-going vessels are described in accordance with the provisions（规定）of the 1982 registration certificate（登记证书）. This provision came into effect in July 1994 and defines a minimum length of 24 meters for sea-going vessels.

1. Fore Perpendicular (F_{pp} or FP)

This line crosses the intersection（交点） of the water line and the front of the stem.

2. Aft Perpendicular (A_{pp} or AP)

This line usually aligns with（重合） the center line of the rudder stock (the imaginary line around which the rudder rotates).

3. Load Line

This is the water line of a ship is floating on water. There are different load lines for different situations, such as

（1）Light Water Line

The water line of a ship which carries only her regular inventory（货物）.

（2）Deep Water Line

The water line of the maximum load draught in seawater.

（3）Design Water Line

The load line indicates（表明） the mark in summer as calculated in ship deign by the ship builder.

4. Deck Side Line

Extended（延伸）line from the topside of the fixed deck at the ship's side.

5. Base Line

Top of the keel.

6. Sheer

This is the upward rise of a ship's deck from amidships（舯）towards the bow and stern. The sheer gives the vessel extra reserve buoyancy（储备浮力） at the stem and the stern.

7. Camber

The deck is arched (拱曲度） in the transverse direction.The curvature helps ensuring sufficient（充分的）drainage.

The above principal dimensions are shown in figure 2.1

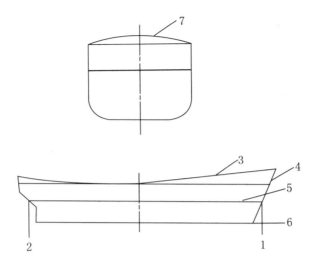

1—Fore Perpendicular; 2—Aft Perpendicular; 3—Design Water Line;
4—Deck Side Line; 5—Base Line; 6—Sheer; 7—Camber.

Fig.2.1　Principal Dimensions

8. Length between Perpendiculars (L_{pp})

Distance between the Fore and the Aft Perpendicular.

9. Length over All (L_{OA})

The horizontal distance from stem to stern.

10. Length on the Water Line (L_{WL})

This is the length of design water-line or the length of full-load waterline.

11. Breadth (B)

The greatest breadth, measured from side to side outside the frames but inside the shell plating.

12. Breadth over All (B_{max})

The maximum breadth of a ship as measured from the outer hull on starboard to the outer hull on port side.

13. Fore Draft (T_f)

It's the vertical distance between the water line and the underside of the keel, as measured on the fore perpendicular.

14. After Draft (T_a)

The vertical distance between the water line and the underside of the keel as measured from the aft perpendicular.

15. Trim

This is the difference between the draught at the stem and that at the stern.

16. Trim by the Bow

It's the condition under which the draft is larger at the stem than that at the stern.

17. Trim by the Stem

It's the condition under which the draft is larger at the stern than that at the stem.

18. Depth

The vertical distance between the base line and the upper continuous deck.

19. Freeboard

The distance between the water line and the top of the deck at the side.

Each of the above main dimensions is shown in figure 2.2

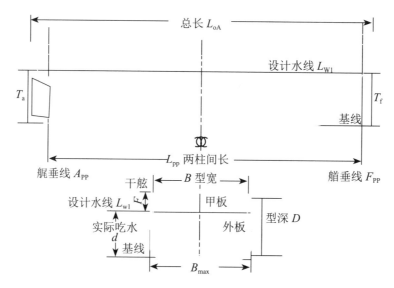

Fig.2.2　Prineipal Dimensions

New Words and Expressions

1. principal ['prɪnsəpl] *adj.* 主要的，首要的

2. dimension [daɪ'menʃn] *n.* 尺寸，尺度

3. perpendicular [ˌpɜːpən'dɪkjələ(r)] *n.* 垂线

4. keel [kiːl] *n.* 龙骨

5. draught [drɑːft] *n.* 吃水

6. topside ['tɒpsaɪd] *n.* 最上层

7. freeboard ['friːˌbɔːd] *n.* 干舷

8. horizontal [ˌhɒrɪ'zɒnt(ə)l] *adj.* 水平的

9. tropical ['trɒpɪk(ə)l] *adj.* 热带的

10. stern [stɜːn] *n.* 艉部；船尾

11. frame [freɪm] *n.* 框架

12. starboard ['stɑ:bəd] *n.* （船舶的）右舷，右侧
13. trim [trɪm] *n.* 纵倾
14. depth [depθ] *n.* 型深
15. sheer [ʃɪə(r)] *n.* 舷弧
16. camber ['kæmbə(r)] *n.* 梁拱
17. curvature ['kɜ:vətʃə(r)] *n.* 曲率
18. ensure [ɪn'ʃuə(r)] *v.* 保证
19. drainage ['dreɪnɪdʒ] *n.* 排水

Notes

1. All sea-going vessels are described in accordance with the provisions（规定）of the 1982 registration certificate. This provision came into effect in July 1994 and defines a minimum length of 24 meters for seagoing vessels.

   译：所有海船外表特征表述依据1982年的登记证书规定。该规定于1994年7月实行，定义的海船最小长度为24米。

2. The maximum breadth of a ship as measured from the outer hull on starboard to the outer hull on port side.

   译：最大船宽是在船体外板左舷到右舷间的距离。

3. It's the vertical distance between the water line and the underside of the keel, as measured on the fore perpendicular.

   译：在艏柱上量取的设计水线与龙骨线间的垂直距离。

Expanding reading

1. Concept Design

The very first effort, concept design, translates the mission requirements into naval architectural and engineering characteristics. It embodies technical feasibility studies to determine such fundamental elements of the proposed ship as length, beam, depth, draft, fullness, power, or alternative sets of characteristics, all of which meet the required speed, range, cargo cubic, and deadweight.

2. Preliminary Design

A ship's preliminary design further refines the major ship characteristics affecting cost and performance. Certain controlling factors such as length, beam, horsepower, and deadweight would not be expected to change upon completion of this phase. Its completion provides a precise definition of a vessel that will meet the mission requirements; this provides the basis for development of contract plans and specifications.

3. Contract Design

The contract design stage yields a set of plans and specifications which form an integral part of the shipbuilding contract document.

4. Detail design

The final stages of ship design is the development of detailed working plans. These plans are the installation and construction instructions to the ship fitters, welders, outfitters, metal workers, machinery vendors, pipe fitters, etc.

New Words and Expressions

1. concept design 概念设计
2. fundamental [ˌfʌndəˈmentl] adj. 基础的 , 基本的
3. preliminary designs 初步设计
4. horsepower [ˈhɔːrspaʊər] n. 马力 (归功率单位)
5. specification [ˌspesɪfɪˈkeɪʃn] n. 规格 , 规范
6. contract design 合同设计
7. integral [ˈɪntɪɡrəl] adj. 必需的 , 不可或缺的
8. detailed design 详细设计
9. installation [ˌɪnstəˈleɪʃn] n. 安装 , 设置
10. machinery vendor 主机供货方

Exercises

Ⅰ. Answer the following questions according to the passage.

1. Can you mark the principal dimensions shown in Fig.2.3?

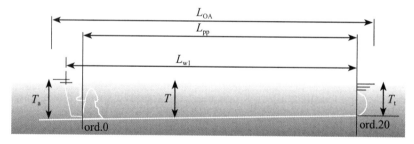

Fig.2.3

2. What types of ships can be classified according to the status of navigation ?

Ⅱ. Practic these new words.

1. English to Chinese.

principal _____   dimension _____

provisions _____   perpendicular _____

stern _____   freeboard _____

2. Chinese to English.

艏垂线 _____   吃水 _____

型宽 _____   型深 _____

主尺度 _____   舷弧 _____

Ⅲ. Explain the nouns.

1. depth

_____

2. freeboard

_____

Ⅳ. Fill in the blanks with the proper words or expressions given below.

> centerline, length, provision, vessels, accordance with, the water line

All sea-going _____ are described in _____ the provisions of the 1982 registration certificate. This _____ came into effect in July 1994 and defines a minimum _____ of 24 meters for sea-going vessels.

1. Fore Perpendicular (F_{PP} or FP)

This line crosses the intersection of _____ and the front of the stem.

2. Aft Perpendicular (A_{PP} or AP)

This line usually aligns with the _____ of the rudder stock (the imaginary line around which the rudder rotates).

Ⅴ. Translation.

1. Translate the following sentences into Chinese.

(1) This line crosses the intersection of the water line and the front of the stem.

_____

(2) The greatest breadth, measured from side to side outside the frames but inside the shell plating.

(3) The water line of a ship which carries only her regular inventory.

(4) The maximum breadth of a ship as measured from the outer hull on starboard to the outer hull on port side.

(5) It's the condition under which the draft is larger at the stem than that at the stern.

2. Translate the short passage.

(1) Load Line

This is the water line of a ship is floating on water. There are different load lines for different situations, such as

① Light Water Line

The water line of a ship which carries only her regular inventory.

② Deep Water Line

The water line of the maximum load draught in seawater.

③ Design Water Line

The load line indicates the mark in summer as calculated in ship deign by the ship builder.

(2) Deck Side Line

Extended line from the topside of the fixed deck at the ship's side.

(3) Base Line

Top of the keel.

(4) Sheer

This is the upward rise of a ship's deck from amidships towards the bow and stern. The sheer gives the vessel extra reserve buoyancy at the stem and the stern.

(5) Camber

The deck is arched in the transverse direction. The curvature helps ensuring sufficient drainage.

Lesson 3　Sea Keeping Performance

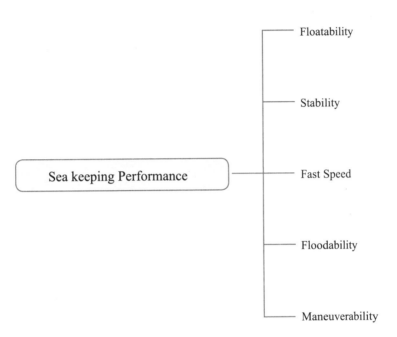

Background

A: Do you know what the sea keeping performance of a ship include?
B: It includes floatability, stability, fast Speed, floodability, maneuverability and so on. Next we will study in further detail.

Text reading

1. Floatability

One of the important ship performances is the floatability that indicates the ability of vessel floating positively on water under the condition of a certain amount of deadweight. When we deal with floatability, must first be clear about the following two technical terms, i. e. reserve buoyancy and load line mark.

Then, what is the reserve buoyancy? As it is known to all, a ship has to obtain a certain amount of freeboard when she's sailing sea. That is to say, to give you a concrete idea, any vessel is to retain some volume above water for the sake of extra buoyancy so that its draft is allowed to increase without a sinking tragedy under particular but rare conditions, such as rough sea or serious flooding due to hull damage. For instance, the film, *TITANIC*, popular in every corner of the world and presenting a moving love story to the sentimental audience, describes the fatal hull damage owing to a huge iceberg. This extra buoyancy is called reserve buoyancy; or, to be exact, the reserve buoyancy, which is measured by freeboard, refers to the watertight volume of hull above load line.In respect of load line mark, it denotes a variety of the max drafts of vessels in different seasons and at varied navigating zones. In China, CCS has worked out *The Rules for Load Line*.

See Figure 3.1 a ship is subjected to gravity and buoyancy in water.Gravity is represented by G, the point of action—center of gravity, buoyancy is represented by B, the action point—center of buoyancy.

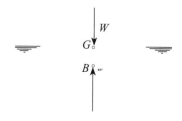

Fig. 3.1　Force and balance of ships

2. Stability

One more ship performance is the stability which is the ability that vessels will incline when affected by an exterior force such as wind wave, etc. ,and will restore its original position on the force removal. Stability is, of course, of great importance to shipbuilding since its failure will always lead to heavy loss of life. History has sadly witnessed tragic sinking of vessels so many a time by reason of poor stability.

Vessel inclination may be divided into trim and heel. As the metacenter of transversal inclination is much more essential than that of longitudinal inclination, the emphases is invariably laid on the transversal stability and discussions are often limited to the small metacentric angle under 15 degrees.

In order to obtain fine stability, precautions will normally be taken in two different ways. On one hand, gravity center is to be lowered; on the other hand, metacenter to be raised. The gravity center of a ship can be calculated by means of an inclination test. The test is generally to be performed in calm water and with lovely weather. If you have got a drydock for the test, it is so much the better. Fig.3.2-Fig.3.4 describe the three equilibrium states of the ship.

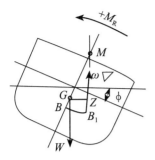
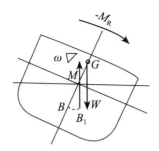
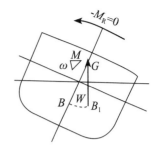

Fig. 3.2 Stable equilibrium   Fig. 3.3 Unstable equilibrium   Fig. 3.4 Neutral equilibrium

3. Fast Speed

Another ship performance is the fast speed or rapidity which describes the ability for a ship to gain faster speed at lower consumption of power.

Vessels will be affected mainly by water resistance during voyage. We do not bother our brain about air resistance because it is far smaller than water resistance. Perhaps, the only exception is high-speed boats.

Water resistance to be encountered by a ship consists of fiction resistance, swirl (eddy-making) resistance and wave-forming (wave-making) resistance.

There are two ways to raise ship speed, namely, to minimize water resistance and to increase main engine power. To do this, a bulbous is widely used in many types of ships and the rated horsepower of main engines is generally to be over two times as big as the effective horsepower of ships.

4. Floodability

Floodability states the ability for a vessel to keep afloat with sufficient floatability, stability and other ship performances in case one or several compartments are flooded. Should sea damage take place, reserve buoyance would be the principal condition to keep a vessel afloat. With the help of the watertight bulkheads and decks which separate the inside of hull into a number of compartments and spaces, reserve buoyance enough is to be retained so that the intake sea water may be confined to the damaged compartment without water pouring into the adjacent compartments.

5. Maneuverability

The last ship performance that we are coming to is maneuverability, which refers to the ability for a vessel to retain or change its course in accordance with the pilot's intention.

Maneuverability is composed of two abilities, that is, the directional stability and the turning ability. The former indicates the ability for a vessel to keep to its given course, while the latter sets forth the ability for a vessel to change its course. Ocean-going vessels require strict directional stability, whereas short-range ships ask for a better turning ability. Moreover, the smaller the turning circle of vessels, the better their turning ability.

In the guarantee of vessel maneuverability, a steering gear of fine quality is to be provider, rudder is its primary component (Fig.3.5).

Fig. 3.5　Turning ability

New Words and Expressions

1. floatability [ˌfləʊtəˈbɪlɪti] *n.* 浮性

2. deadweight 载重量

3. reserve [rɪˈzɜːv] *n.* 储备；保护区；保留

4. buoyancy [ˈbɔɪənsi] *n.* 浮力；弹性

5. sinking [ˈsɪŋkɪŋ] *n.* 沉没

6. rough sea 汹涌的大海

7. sentimental [ˌsentɪˈmentl] *adj.* 伤感的；多愁善感的

8. fatal [ˈfeɪtl] *adj.* 致命的；毁灭性的；决定性的

9. stability [stəˈbɪləti] *n.* 稳定性；坚定，恒心

10. exterior [ɪkˈstɪəriə(r)] *adj.* 外部的；表面的；外在的

11. inclination [ˌɪnklɪˈneɪʃn] *n.* 倾向，倾斜

12. metacenter [ˈmetəsentə(r)] *n.* 稳心

13. essential [ɪˈsenʃl] *adj.* 基本的；必要的；本质的

14. fast speed 快速性

15. consumption [kənˈsʌmpʃn] *n.* 消费；消耗

16. resistance [rɪˈzɪstəns] *n.* 抵抗；阻力；抗力

17. floodability 抗沉性

18. adjacent [əˈdʒeɪs(ə)nt] *adj.* 邻近的；毗连的；接近的

19. maneuverability [məˌnuvəˈbɪlɪtɪ] *n.* 可操作性；机动性

20. guarantee [ˌɡærənˈtiː] *n.* 保证，担保

Notes

1. One of the important ship performances is the floatability that indicates the ability of vessel floating positively on water under the condition of a certain amount of deadweight.

   译：浮性是船舶性能的重要指标之一，它反映了船舶在一定载重量条件下正浮于水面的能力。

2. This extra buoyancy is called reserve buoyancy; or, to be exact, the reserve buoyancy, which is measured by freeboard, refers to the watertight volume of hull above load line.

   译：这种额外的浮力称为储备浮力；或者确切地说，用干舷测量的储备浮力是指船体在载重线上的水密体积。

3. One more ship performance is the stability which is the ability that vessels will incline when affected by an exterior force such as wind wave, etc., and will restore its original position on the force removal.

   译：船舶性能的另一个重要指标是稳性，即船舶在受到外力（如风浪等）影响时倾斜，并在撤销外力后，恢复其原位的能力。

Expanding reading

Roll and Pitch

When floating on water or sailing at sea vessels will roll or pitch owing to wave motions as well as the influence of wind, current and propeller.

The results of excessive roll and pitch are as follows:

A. First, the ship upsetting arising from excessive inclination due to roll.

B. In the next place, the hull structure damage because of sharp roll and pitch as well as the violent movement of bulk cargoes, or even worse the hull might break.

C. Still next, the affection on the propulsion plant, i. e. the increase of water resistance and the reduction of speed by reason of roll and pitch.

D. Then, the affection on proper operation of various kinds of machines and instruments.

E. Last of all, the hard working condition which causes the crew to be seasick.

Therefore, we have to take roll and pitch into our consideration during the design stage, for they are closely related to the whole ship performance.

As you know, the roll period is greatly concerned with the initial metacentric height and, in some degree, stability is contrary to roll. It seams to be strange that violent roll, as many people might think, is not derived from poor stability.

So far different stabilizing units have been invented, and found wide application to decreasing roll and pitch（Fig.3.6）. Common stabilizing units now in practical use are bilge keels, stabilizers and water ballast tanks.

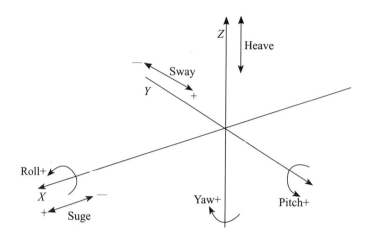

Fig. 3.6   Roll and Pitch

New Words and Expressions

1. roll an pitch 横摇和纵摇
2. propeller [prəˈpelər] *n.* 螺旋桨（飞机或轮船的推进器）
3. upsetting arising 倾覆
4. excessive inclination 过度倾斜
5. seasick [ˈsiːsɪk] *adj.* 晕船
6. metacentric [ˈmetəsentrɪk] *adj.* 定倾中心的，稳定中心的
7. derived [dɪˈraɪvd] *v.*(使)起源，(使)产生
8. water ballast tanks 压载水舱

Exercises

I. Answer the following questions according to the passage.

1. Can you tell me what the sea keeping performance of a ship include?

_____

2. What's the mean of maneuverability?

_____

II. Practic these new words.

1. English to Chinese.

   deadweight _____   sinking _____

rough sea _____   consumption _____

resistance _____   guarantee _____

2. Chinese to English.

浮性 _____   稳心 _____

快速性 _____   抗沉性 _____

操纵性 _____   稳性 _____

Ⅲ. Explain the nouns.

1. floatability

_____

2. floodability

_____

Ⅳ. Fill in the blanks with the proper words or expressions given below.

> essential, of course, lead to, metacenter, exterior, inclination, transversal

One more ship performance is the stability which is the ability that vessels will incline when affected by an _____ force such as wind wave, etc. ,and will restore its original position on the force removal. Stability is, _____ , of great importance to shipbuilding since its failure will always _____ heavy loss of life. History has sadly witnessed tragic sinking of vessels so many a time by reason of poor stability.

Vessel _____ may be divided into trim and heel. As the _____ of transversal inclination is much more _____ than that of longitudinal inclination, the emphases is invariably laid on the _____ stability and discussions are often limited to the small metacentric angle under 15 degrees.

Ⅴ. Translation.

1. Translate the following sentences into Chinese.

(1) In order to obtain fine stability, precautions will normally be taken in two different ways.

_____

（2）If you have got a drydock for the test, it is so much the better. Fig. 3.2-Fig. 3.4 describe the three equilibrium states of the ship.

（3）Vessels will be affected mainly by water resistance during voyage. We do not bother our brain about air resistance because it is far smaller than water resistance.

（4）Water resistance to be encountered by a ship consists of fiction resistance, swirl (eddy-making) resistance and wave-forming (wave-making) resistance.

（5）There are two ways to raise ship speed, namely, to minimize water resistance and to increase main engine power.

2. Translate the short passage.

（1）Floatability

One of the important ship performances is the floatability that indicates the ability of vessel floating positively on water under the condition of a certain amount of deadweight. When we deal with floatability, must first be clear about the following two technical terms, i.e. reserve buoyancy and load line mark.

Then, what is the reserve buoyancy? As it is known to all, a ship has to obtain a certain amount of freeboard when she's sailing sea. That is to say, to give you a concrete idea, any vessel is to retain some volume above water for the sake of extra buoyancy so that its draft is allowed to increase without a sinking tragedy under particular but rare conditions, such as rough sea or serious flooding due to hull damage. For instance, the film, *TITANIC*, popular in every corner of the world and presenting a moving love story to the sentimental audience, describes the fatal hull damage owing to a huge iceberg. This extra buoyancy is called reserve buoyancy; or, to be exact, the reserve buoyancy, which is measured by freeboard, refers to the watertight volume of hull above load line. In respect of load line mark, it denotes a variety of the max. drafts of vessels in different seasons and at varied navigating zones. In China, CCS has worked out *The Rules for Load Line*.

A ship is subjected to gravity and buoyancy in water. Gravity is represented by G, the point of action-center of gravity, buoyancy is represented by B, the action point--center of buoyancy.

Module 3　Ship Structure

Lesson 4　Equipment to Ensure Stability

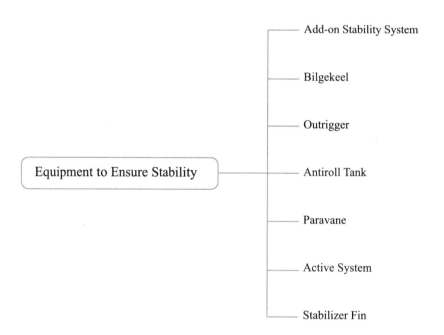

Background

A: Do you know what the equipment to ensure stability include?

B: It includes add-on stability systems, bilge keel, outriggers, antiroll tanks and so on. Next we will study in further detail.

Text reading

1. Add-on Stability System

　　These systems are designed to reduce the effects of waves or wind gusts. They do not increase the stability of the vessel in a calm sea. The International Maritime Organization *International Convention on Load Lines* does not mention active stability systems as a method of ensuring stability. The hull must be stable without active systems.

2. Bilge Keel

　　A bilge keel(Fig.4.1)is a long fin of metal,often in a "V" shape. A bilge keel is a long board located along the outer side of the bilge near the middle of the ship. It is along the direction of the current and has a length of about 1/4 to 1/3 the length of the ship.It is used to reduce the ship's roll.

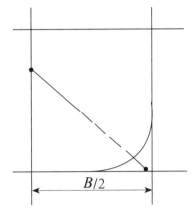

Fig. 4.1 Bilge keel   Fig. 4.2 Position of the bilge keel

Bilge keels are employed in pairs (one for each side of the ship). A ship may have more than one bilge keel per side, but this is rare. Bilge keels increase the hydrodynamic resistance when a vessel rolls, thus limiting the amount of roll a vessel has to endure.In the cross-section direction, the bilge keel is approximately perpendicular to the bilge strake, and its outer edge cannot exceed the area enclosed by the baseline of the bottom of the ship and the sidelines to avoid damage when approaching the dock (Fig.4.2).

3. Outrigger

Outriggers may be employed on certain vessels to reduce rolling. Rolling is reduced either by the force required to submerge buoyant floats or by hydrodynamic foils. In some cases these outriggers may be of sufficient size to classify the vessel as a trimaran; however on other vessels they may simply be referred to as stabilizers (Fig.4.3).

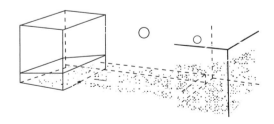

Fig. 4.3 Outriggers   Fig. 4.4 Antiroll tanks

4. Antiroll Tank

Antiroll tanks are tanks within the vessel fitted with baffles intended to slow the rate of water transfer from the port side of the tank to the starboard side. The tank is designed such that a larger amount of water is trapped on the higher side of the vessel. This is intended to have an effect completely opposite to that of the free surface effect (Fig.4.4).

5. Paravanes

A device equipped with sharp teeth and towed alongside a ship to cut the mooring cables of submerged mines.Paravanes may be employed by slow-moving vessels (such as fishing vessels) to reduce roll (Fig.4.5).

Fig. 4.5 Paravanes

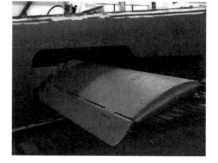

Fig. 4.6 Stabilizer fins

6. Active System

Many vessels are fitted with active stability systems. Active stability systems are defined by the need to input energy to the system in the form of a pump, hydraulic piston, or electric actuator. These systems include stabilizer fins attached to the side of the vessel or tanks in which fluid is pumped around to counteract the motion of the vessel.

7. Stabilizer Fin

Active fin stabilizers are normally used to reduce the roll that a vessel experiences while underway or, more recently, while at rest. The fins extend beyond the hull of the vessel below the waterline and alter their angle of attack depending upon heel angle and rate-of-roll of the vessel. They operate similar to airplane ailerons. Cruise ships and yachts frequently use this type of stabilizer system.

When fins are not retractable, they constitute fixed appendages to the hull, possibly extending the beam or draft envelope, requiring attention for additional hull clearances (Fig.4.6).

New Words and Expressions

1. add-on stability systems 附加稳性系统
2. bilge keel 舭龙骨
3. board [bɔːd] n. 木板；船舷或机舱
4. hydrodynamic [ˌhaɪdrəʊdaɪˈnæmɪk] adj. 流体动力学的；水力的
5. resistance [rɪˈzɪstəns] n. 抵抗；阻力；抗力
6. cross-section direction 横剖面

7. approximately [əˈprɒksɪmətli] *adv.* 大约；近似地
8. bilge strake 舭列板
9. outrigger [ˈaʊtrɪɡə(r)] *n.* 舷外支架
10. trimaran [ˈtraɪməræn] *n.* 三体帆船
11. antiroll tank 减摇水舱
12. baffle [ˈbæfl] *n.* 隔板，挡板；遮护物
13. paravane [ˈpærə,veɪn] *n.* 平准翼板
14. mooring [ˈmɔːrɪŋ] *n.* 下锚；系船用具；停泊处
15. active system 主动系统
16. hydraulic piston 液压活塞
17. electric actuator 电动装置
18. stabilizer fin 减摇鳍
19. aileron [ˈeɪlərɒn] *n.* 副翼
20. beam [biːm] *n.* 横梁

Notes

1. Outriggers may be employed on certain vessels to reduce rolling. Rolling is reduced either by the force required to submerge buoyant floats or by hydrodynamic foils.
   译：可以在某些船上使用支腿支架以减少摇摆。淹没浮标所需的力或流体动力可以有效减少摇摆。

2. Antiroll tanks are tanks within the vessel fitted with baffles intended to slow the rate of water transfer from the port side of the tank to the starboard side.
   译：减摇舱是船内装有挡板的舱，其目的是减缓水从舱的左舷向右舷的转移速度。

3. A device equipped with sharp teeth and towed alongside a ship to cut the mooring cables of submerged mines. Paravanes may be employed by slow-moving vessels (such as fishing vessels) to reduce roll.
   译：一种装置，它装有锋利的牙齿，与船一起拖曳，用于切断水下地雷的系泊电缆。缓慢移动的船只（例如渔船）可以使用平准翼板来减少侧倾。

Expanding reading

Gyroscopic Stabilizer

Gyroscopic stabilizer is a piece of naval equipment that counters the chaotic whirlpools threat in naval garrison missions. Gyroscopes were first used to control a ship's roll in the late 1920s and early 1930s for warships and then passenger liners. The most ambitious use of large gyroscope to control a ship's roll was on an Italian passenger liner, the *SS Conte di Savoia*, in which three large gyroscope were mounted in the forward part of the ship. While they proved successful in drastically reducing roll in the westbound trips, the system had to be disconnected on the eastbound leg for safety reasons. This was because with a following sea (and the deep slow rolls this generated) the vessel tended to "hang" with the system turned on, and the inertia it generated made it harder for the vessel to right itself from heavy rolls.

When the boat rolls, the rotation acts as an input to the gyroscope, causing the gyroscope to generate, rotation around its output axis such that the spin axis rotates to align itself with the input axis. This output rotation is called precession and, in the boat case, the gyroscope will rotate fore and aft about the output or gimbal axis（Fig.4.7）.

Fig. 4.7 Gyroscopic Stabilizer

New Words and Expressions

1. gyroscopic stabilizer 陀螺稳定器
2. chaotic whirlpools threat 混乱漩涡威胁
3. drastically ['dræstɪkli] *adv.* 大大地，彻底地
4. the SS Conte di Savoia 萨伏伊伯爵号
5. inertia [ɪ'nɜːʃə] *n.* 惰性，惯性
6. output axis 输出轴
7. spin axis 自转轴
8. precession [prɪ'seʃən] 进动性，旋进

Exercises

I. Answer the following questions according to the passage.

1. Can you tell me what the equipment to ensure stability include?

2. What's the function of antiroll tanks?

Lesson 4 Equipment to Ensure Stability

II. Practic these new words.

1. English to Chinese.

   stabilizer fin _____   active system _____

   paravane _____   stabilizer fin _____

   trimaran _____   outrigger _____

2. Chinese to English.

   舭龙骨 _____   横剖面 _____

   舭列板 _____   减摇水舱 _____

   液压活塞 _____   电动装置 _____

III. Explain the nouns.

1. bilge keel

   _____

2. stabilizer fins

   _____

IV. Fill in the blanks with the proper words or expressions given below.

> baseline, direction, board, bilge strake,
> cross-section, hydrodynamic resistance, bilge keel

A _____ is a long fin of metal, often in a "V" shape. A bilge keel is a long _____ located along the outer side of the bilge near the middle of the ship. It is along the _____ of the current and has a length of about 1/4 to 1/3 the length of the ship. It is used to reduce the ship's roll.

Bilge keels are employed in pairs (one for each side of the ship). A ship may have more than one bilge keel per side, but this is rare. Bilge keels increase the _____ when a vessel rolls, thus limiting the amount of roll a vessel has to endure. In the _____ direction, the bilge keel is approximately perpendicular to the _____, and its outer edge cannot exceed the area enclosed by the _____ of the bottom of the ship and the sidelines to avoid damage when approaching the dock.

Ⅴ. Translation.

1. Translate the following sentences into Chinese.

(1) In some cases these outriggers may be of sufficient size to classify the vessel as a trimaran; however on other vessels they may simply be referred to as stabilizers.

(2) The tank is designed such that a larger amount of water is trapped on the higher side of the vessel.

(3) Paravanes may be employed by slow-moving vessels (such as fishing vessels) to reduce roll.

(4) These systems include stabilizer fins attached to the side of the vessel or tanks in which fluid is pumped around to counteract the motion of the vessel.

(5) The fins extend beyond the hull of the vessel below the waterline and alter their angle of attack depending upon heel angle and rate-of-roll of the vessel.

2. Translate the short passage.

1. Add-on Stability Systems

These systems are designed to reduce the effects of waves or wind gusts. They do not increase the stability of the vessel in a calm sea. The International Maritime Organization *International Convention on Load Lines* does not mention active stability systems as a method of ensuring stability. The hull must be stable without active systems.

2. Bilge Keel

A bilge keel is a long fin of metal, often in a "V" shape. A bilge keel is a long board located along the outer side of the bilge near the middle of the ship. It is along the direction of the current and has a length of about 1/4 to 1/3 the length of the ship. It is used to reduce the ship's roll.

Bilge keels are employed in pairs (one for each side of the ship). A ship may have more than one bilge keel per side, but this is rare. Bilge keels increase the hydrodynamic resistance when a vessel rolls, thus limiting the amount of roll a vessel has to endure. In the cross-section direction, the bilge keel is approximately perpendicular to the bilge strake, and its outer edge cannot exceed the area enclosed by the baseline of the bottom of the ship and the sidelines to avoid damage when approaching the dock.

Lesson 5　Ship Structure (1)

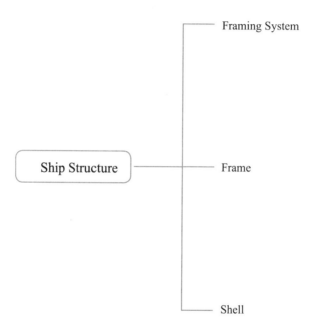

Background

A: Do you know what the inside of the hull structure looks like?

B: In general, the inside of a ship is composed of many crisscross framing structures, which can ensure the structural strength of the ship. Next we will study in further detail.

Text reading

A ship is something like a grand mansion floating on water and with a number of "floors" called decks. The hull can be roughly divided into two parts, the main hull (Fig.5.1) and the superstructure. The main hull refers to the part below the upper deck and the superstructure refers to the part above the upper deck. The main hull is a watertight hollow structure surrounded by ship bottom, ship side and upper deck. However, if you want to learn more of the hull construction, then you must go into the details of primary structures of hull.

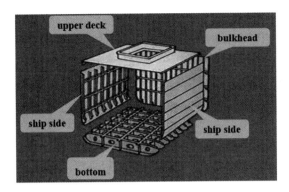

Fig. 5.1　The main hull

1. Framing System

According to the direction of framing arrangement, the framing system can be divided into three types: transverse system, longitudinal system and combined system.

(1) transverse framing system

A large number of closely spaced frames are arranged along the breadth (transverse) direction of the ship. These transverse frames are further supported by widely spaced longitudinal members. The advantage is that most of the frames are arranged transversely, the transverse strength is better, the construction is more convenient, and the construction cost is low; the disadvantage is that under the same load, the thickness of the outer plate and deck is larger than that of the longitudinal system, and the structure weight is larger.

(2) longitudinal framing system

A large number of closely spaced frames are arranged along the ship's longitudinal direction. The longitudinal system of framing differs from the above in that the shell and bottom plating are supported by closely spaced members running fore and aft. These longitudinal members are then reinforced by widely space transverse frames. Its advantage is that most of the frames are longitudinally arranged, which increases the effective area of the ship beam to resist longitudinal bending, increases the longitudinal bending capacity of the ship beam, and increases the total longitudinal strength of the hull; the disadvantage is that the construction is relatively troublesome.

(3) combined framing system

The frames in the transverse and longitudinal directions are similar, and the spacings are nearly equal. This framing system is rarely used except for special occasions.

The type of framing system used is related to the type of cargo carried. For example, tankers are always constructed with a longitudinal system since tremendous transverse members are needed in this system to stiffen the vessel transversely so as not to interfere the carriage of liquids. Freighters carrying general cargo cannot use this system since these transverse members would impede the stowage of cargo (Fig.5.2).

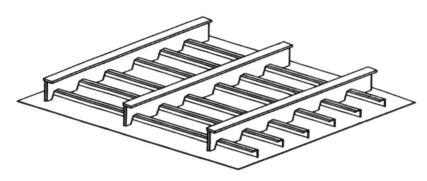

Fig. 5.2 Framing System

2. Frame

The most important function of frames is to stiffen the shell plating and compel it to retain its proper form against the pressure of the water outside. The shell plating alone is somewhat flexible and without the support of the frames, it would collapse. The frames also transfer weights from the deck to the bottom structure, where they are neutralized by the ship's buoyancy.

Frames may be formed of a simple shape, or they may be built up of plate and angles. The simple shapes are used for side frames; built-up frames are found in the bottom structure. The deepened transverse frames across the bottom of the hull are known as the floors.

3. Shell

One of the basic structures in hull construction is the shell structure which ensures vessel floatability. It is the water-tight shell structure that is subjected to different external forces, such as the total longitudinal bend, water pressure, wave impact, squeeze of ice blocks, etc.

In the shell structures, each shell plating has its own designation. For example, the plating located at the ship bottom is called bottom plates, among which are keel plates that are a series of plating sited at the center line of hull. Bilge strakes denote the plating at the bending areas from bottom to side. And the plating distributed at both sides is termed as side plates. Among them are topside plates that are in connection with the upper deck (Fig. 5.3).

Shell plating stands external forces of different magnitude in the light of its various situations. Accordingly, the thickness of plates will vary with the force to be subjected and there's no doubt that, at some key zones, local strengthening is inevitable.

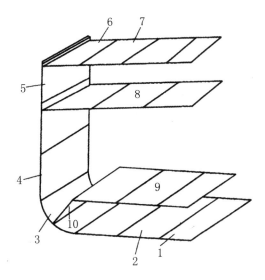

1—keel plate；2—bottom plate；3—bilge strake；4—side plate；5—topside plate；
6—deck stringer plate；7—upperdeck；8—lower deck；9—inner bottom plate；10—margin plate.

Fig. 5.3 Designation of outer and deck plating

New Words and Expressions

1. the main hull 主船体
2. superstructure[ˈsuːpəstrʌktʃə(r)] n. (船舶的) 上层建筑
3. the upper deck 上甲板
4. watertight[ˈwɔːtətaɪt] adj. 水密的
5. construction[kənˈstrʌkʃn] n. 建造
6. framing system 骨架形式
7. transverse[ˈtrænzvɜːs] adj. 横 (向) 的
8. longitudinal[ˌlɒŋgɪˈtjuːdɪnl] adj. 纵的；纵向的
9. the total longitudinal strength 总纵强度
10. stringer[ˈstrɪŋə(r)] n. 桁材
11. aggregate[ˈægrɪgə] n. 骨材
12. stiffen[ˈstɪf(ə)n] v. (使) 变强硬，变坚定
13. collapse[kəˈlæps] v. (突然) 倒塌，坍塌
14. squeeze [skwiːz] v. 挤压
15. magnitude [ˈmægnɪtjuːd] n. 等级
16. keel plate 平板龙骨
17. bilge strake 舭列板
18. topside plate 舷顶列板
19. deck stringer plate 甲板边板
20. margin plate 内底边板

Notes

1. The main hull refers to the part below the upper deck and the superstructure refers to the part above the upper deck. The main hull is a watertight hollow structure surrounded by ship bottom, ship side and upper deck.

    译：船体大致可分为主船体和上层建筑两部分，主船体指上甲板以下部分，上层建筑指上甲板以上部分。 主船体是由船底、舷侧、上甲板围成的水密的空心结构。

2. According to the direction of framing arrangement, the framing system can be divided into three types: transverse system, longitudinal system and combined system.

    译：板架结构根据骨材布置的方向，可分为纵骨架式、横骨架式和混合骨架式三种类型。

3. The most important function of frames is to stiffen the shell plating and compel it to retain its proper form against the pressure of the water outside.

    译：骨架最重要的功能是加强外壳板并迫使其保持适当的形状，以抵抗外部水的压力。

4. It is the water-tight shell structure that is subjected to different external forces, such as the total longitudinal bend, water pressure, wave impact, squeeze of ice blocks, etc.

    译：水密壳结构承受不同的外力，例如总纵弯曲、水压力、波浪冲击力、冰块的挤压等。

Expanding reading

The joint of outer plate

The long side of the outer plate is usually arranged along the longitudinal direction of the ship. The longitudinal joint between the plates is called the side seam, and the transverse joint between the plats is called the butt, as shown in Fig.5.4.

When arranging the side seam lines, the arrangement of the longitudinal members of the hull shall be considered. The side seam of the outer plate and the fillet weld of the longitudinal members shall not overlap or form too small intersection angle, otherwise the welding quality will be affected. If the intersection angle of the longitudinal member and the joint of the outer plate edge is less than 30°, the joint should be adjusted to a stepped shape, as shown in Fig.5.5. In addition, when the plate seam arrangement is parallel to the longitudinal member over a long distance, the distance between them should be greater than 50 mm.

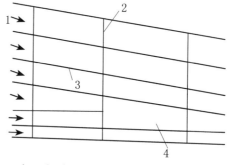

1—strake; 2—butt; 3—side seam; 4—stealer strake.

Fig. 5.4 The joint

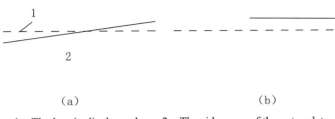

(a)　　　　　　　　　　(b)

1—The longitudinal member; 2—The side seam of the outer plate

Fig. 5.5　The side seam of the outer plate

Generally, there are two types of steeler strakes: one is a double steeler strake, and the end seams of two adjacent rows of plates are interrupted at the same time to form a row of plates; the other is lamina strake, the end seams of two adjacent rows of plates are not interrupted at the same time, stepped seams are formed at the steeler strakes, as shown in Fig.5.6.

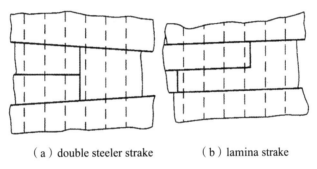

(a) double steeler strake　　(b) lamina strake

Fig. 5.6　The steeler strakes of the outer plate

New Words and Expressions

1. butt [bʌt] *n.* 端接缝
2. side seam 边接缝
3. overlap [ˌəʊvəˈlæp] *v.* 部分重叠,交叠
4. intersection [ˌɪntəˈsekʃn] *n.* 交点,相交
5. parallel [ˈpærəlel] *adj.* 平行
6. double steeler strake 双并板
7. lamina strake 齿形并板
8. adjacent [əˈdʒeɪsnt] *adj.* 与……毗连的,邻近的

Exercises

I. Answer the following questions according to the passage.

1. What does the watertight hollow structure consist of ?

_____

2. Do you know the sign of outer and deck plating?

_____

II. Practic these new words.

1. English to Chinese.

  keel plate _____   bilge strake _____

  topside plate _____   deck stringer plate _____

  margin plate _____   watertigh _____

2. Chinese to English.

  主船体 _____   上层建筑 _____

  上甲板 _____   骨架形式 _____

  总纵强度 _____   桁材 _____

III. Explain the nouns.

1. the main hull

_____

2. keel plate

_____

IV. Fill in the blanks with the proper words or expressions given below.

> beam, to resist, fore and aft, from,
> advantage, large number of, strength

  A _____ closely spaced frames are arranged along the ship's longitudinal direction. The longitudinal system of framing differs _____ the above in that the shell and bottom plating are supported by closely spaced members running _____. These longitudinal members are then reinforced by widely space transverse frames. Its _____ is that most of the frames are longitudinally arranged, which increases the effective area of the ship beam _____ longitudinal bending, increases the longitudinal bending capacity of the ship _____, and increases the total longitudinal _____ of the hull; the disadvantage is that the construction is relatively troublesome.

V. Translation.

1. Translate the following sentences into Chinese.

(1) A ship is something like a grand mansion floating on water and with a number of "floors" called decks. The hull can be roughly divided into two parts, the main hull and the superstructure.

_____

(2) The disadvantage is that under the same load, the thickness of the outer plate and deck is larger than that of the longitudinal system, and the structure weight is larger.

_____

(3) The frames in the transverse and longitudinal directions are similar, and the spacings are nearly equal. This framing system is rarely used except for special occasions.

_____

(4) Freighters carrying general cargo cannot use this system since these transverse members would impede the stowage of cargo.

_____

(5) The shell plating alone is somewhat flexible and without the support of the frames, it would collapse.

_____

2. Translate the short passage.

One of the basic structures in hull construction is the shell structure which ensures vessel floatability. It is the water-tight shell structure that is subjected to different external forces, such as the total longitudinal bend, water pressure, wave impact, squeeze of blocks of ice, etc.

In the shell structures, each shell plating has its own designation. For example, the plating located at the ship bottom is called bottom plates, among which are keel plates that are a series of plating sited at the center line of hull. Bilge strakes denote the plating at the bending areas from bottom to side. And the plating distributed at both sides is termed as side plates. Among them are topside plates that are in connection with the upper deck.

Shell plating stands external forces of differen magnitude in the light of its various situations. Accordingly, the thickness of plates will vary with the force to be subjected and there's no doubt that, at some key zones, local strengthening is inevitable.

Lesson 6　Ship Structure (2)

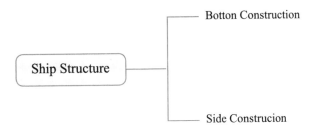

Background

A: Do you know what the inside of the hull construction looks like?

B: In general, the inside of a ship is composed of many crisscross framing constructions, which can ensure the structural strength of the ship. Next we will study in further detail.

Text reading

In addition to shell construction, there are bottom construction, side construction, deck construction, bulkhead construction, bow and stern construction and super construction, too.

1. Bottom Construction

Ship bottom will normally be in two different patterns, single bottom and double bottom. According to the direction of framing arrangement, ship bottom can be divided into two types: transversely framed bottom construction, longitudinal framed bottom construction.

（1）Characteristics of transversely framed single bottom construction

The composition of transversely framed single bottom construction is composed of bottom plating, center keelson, side keelson and floor, as shown in Fig.6.1. It is characterized by simple construction and convenient construction, which is mainly used for small ships such as tugboats, fishing boats and river boats.

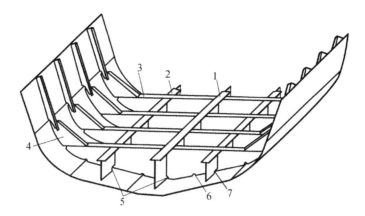

1—center keelson; 2—side keelson; 3—floor; 4—bilge bracket;
5—welding incision; 6—drain hole; 7—bottom plating.

Fig. 6.1 Transversely Framed Single Bottom

(2) Characteristics of longitudinal framed single bottom construction

The longitudinal framed single bottom construction is composed of bottom plating, keelson, floor and a large number of bottom longitudinal, as shown in Fig.6.2. It has good longitudinal strength and light weight, but the process is more complicated, and it is often found in small ships.

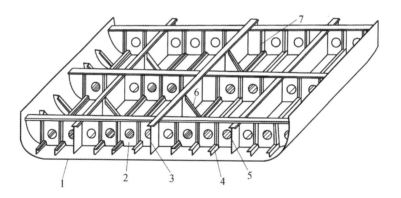

1—bottom plating; 2—floor; 3—center keelson; 4—bottom longitudinal;
5—side keelson; 6—bracket; 7—stiffener.

Fig. 6.2 Longitudinal Framed Single Bottom

(3) Characteristics of transversely framed double bottom construction:

This construction is composed of bottom plating, inner bottom plating, center girder, side girder and various forms of floors (Fig.6.3).

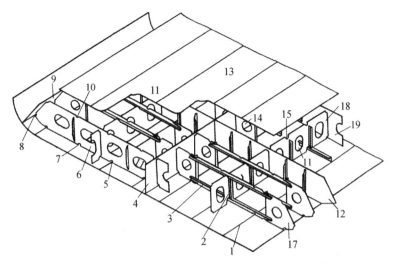

1—side seam; 2—stiffener; 3—bottom frame; 4—center girder; 5—drain hole; 6—solid floor; 7—stiffener rib; 8—welding incision; 9—margin plate; 10—air hole; 11—man hole; 12—bracket; 13—inner bottom plating; 14—lightening hole; 15—incision; 16—inner bottom frame; 17—bracket floor; 18—side girder; 19—solid floor.

Fig. 6.3 Transversely Framed Double Bottom

(4) Characteristics of longitudinal framed double bottom construction

This construction is composed of bottom plating, inner bottom plating, bottom longitudinal, floor, bottom girder. The inner and outer bottom platings are supported by dense longitudinal frames, which increase the rigidity and stability of the plate and increase the longitudinal strength of the bottom. Therefore, the inner and outer bottom platings of longitudinal framed construction can be thinner than the transversely framed construction, which can reduce the structural weight. This kind of framed construction is widely used in modern large and medium-sized ships. (Fig.6.4)

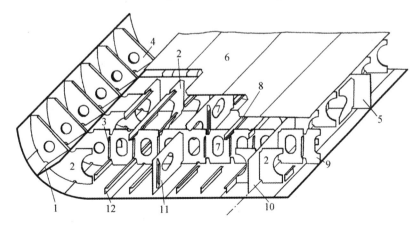

1—margin plate; 2—bracket; 3—stiffener rib; 4—bilge bracket; 5—watertight floor; 6—inner bottom plating; 7—man hole; 8—inner bottom longitudinal; 9—solid floor; 10—center girder; 11—side girder; 12—bottom longitudinal.

Fig. 6.4 Longitudinal Framed Double Bottom

The double bottom is, in fact, watertight compartment. In the double bottom, the center girder is a longitudinal member located at the center line of bottom from bow to stern, which assures you of the total longitudinal strength of hull. Side girders are also longitudinal members situated at both sides of the center line of bottom, which mainly withstand the total longitudinal bend. Floors play the part of chief transversal members at the bottom, which are subjected to water pressure and gravity of cargoes & equipment as well as local external forces. They shoulder the responsibility of transversal and local strength of bottom.

2. Side Construction

Side constructions are subdivided into two models, i. e. the transversal framing and the longitudinal framing. As it were, the side constructions port and starboard are two side walls of hull symmetrical to each other, and among the side constructions leading components are the frames that bear water pressure and secure the transversal and local strength of hull.

There are three forms of side construction:

(1) The form of only one kind frame (Fig.6.5)

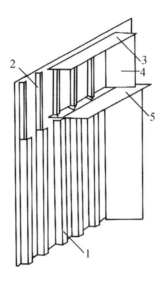

1—main frame; 2—tweendeck frame; 3—upper deck; 4—transverse bulkhead; 5—lower deck.

Fig.6.5　The form of only one kind frame

（2）The form of web frame side stringer and main frame（Fig.6.6）

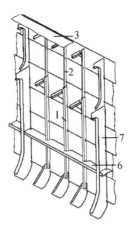

1—bilge bracket；2—main frame；3—bracket；4—beam；5—tweendeck frame；6—web beam；7—web frame；8—side stringer.

Fig.6.6　The form of web frame.side stringer and main frame

（3）The form of double side shell（Fig.6.7）

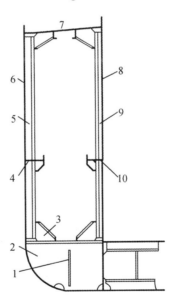

1—stiffener rid；2—lightened floor；3—bracket；4—side girder；5—main frame；6—side shell；7—deck；8—longitudinal bulkhead；9—vertical stiffener；10—horizontal girder.

Fig.6.7　The form of double side shell

New Words and Expressions

1. stiffener [ˈstɪfənə(r)] *n.* 扶强材
2. center keelson 中内龙骨

3. bilge bracket 舭肘板

4. welding incision 焊缝切口

5. side seam 边接缝

6. center girder 中底桁

7. solid floor 实肋板，主肋板

8. stiffener rib 加强筋

9. margin plate 内底边板

10. bracket floor 框架肋板

11. rigidity [rɪˈdʒɪdəti] *n.* 刚性

12. bottom longitudinal 底纵骨

13. transversely [trænsˈvɜːrsli] *adv.* 横向地

14. longitudinal [ˌlɑːndʒəˈtuːdənl] *adj.* 纵向的

15. port [pɔːt] *n.* 左舷

16. symmetrical [sɪˈmetrɪkl] *adj.* 对称的

17. web beam 强横梁

18. side stringer 舷侧纵桁

19. horizontal girder 水平桁

Notes

1. According to the direction of framing arrangement, ship bottom can be divided into two types: transversely framed bottom construction, longitudinal framed bottom construction.

   译：根据骨架布置的方向，船底可分为两种：横骨架式底部结构和纵骨架式底部结构。

2. It is characterized by simple construction and convenient construction, which is mainly used for small ships such as tugboats, fishing boats and river boats.

   译：横骨架式单层底结构的特点是结构简单、建造方便，主要用于拖船、渔船、内河船等小型船舶上。

3. It has good longitudinal strength and light weight, but the process is more complicated, and it is often found in small ships.

   译：纵骨架式单层底结构纵向强度好、结构质量轻，但工艺较复杂，常见于小型舰艇等。

Expanding reading

Aft Ship

The most eye-catching aft spaces on most ships are the engine room and the accommodation. Besides, there can also be working places, storage facilities and fuel or ballast tasks. The aft peak is the part of the ship that is enclosed by the aft peak bulkhead, the stern and the aft deck. The aft peak is the location through which the main engine shaft runs. For support there are floors in the aft peak. The stern construction endures water pressure, vibration resulting from propeller operation and wave impact stirred up by propeller(Fig.6.8).

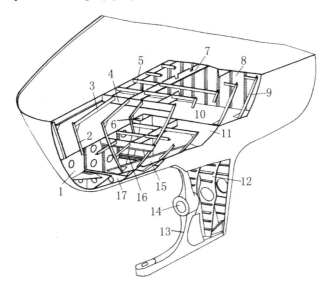

1—wash bulkhead; 2—cant frame; 3—cant beam; 4—web beam; 5—beam; 6—rudder case; 7—deck girder; 8—transverse bulkhead; 9—frame; 10—steering gear room platform; 11—after peak bulkhead; 12—after raised floor; 13—stern post; 14—hub; 15—side stringer; 16—panting beam; 17—floor.

Fig.6.8　Aft Ship

New Words and Expressions

1. wash bulkhead 制荡舱壁
2. cant frame 斜肋骨
3. cant beam 斜横梁
4. rudder case 舵杆管
5. steering gear room platform 舵机舱平台
6. after peak bulkhead 艉尖舱壁
7. after raised floor 艉升高肋板
8. stern post 艉柱
9. hub [hʌb] *n.* 轴毂
10. panting beam 强胸横梁

Exercises

I. Answer the following questions according to the passage.

1. What could the ship bottom be divided into by the direction of framing arrangement?

2. What is the advantage of transversely framed bottom construction?

___

Ⅱ. Practic these new words.

1. English to Chinese.

   stiffener _____   side seam _____

   center keelson _____   bilge bracket _____

   bracket floor _____   bottom longitudinal _____

2. Chinese to English.

   水平桁 _____   内底边板 _____

   强横梁 _____   舷侧纵桁 _____

   中底桁 _____   加强筋 _____

Ⅲ. Explain the nouns.

1. center girder

___

2. bottom longitudinal

___

Ⅳ. Fill in the blanks with the proper words or expressions given below.

> as well as, center line, longitudinal, water pressure, central girder, in fact, side girders

The double bottom is, _____, watertight compartment. In the double bottom, the _____ is a _____ member located at the center line of bottom from bow to stern, which assures you of the total longitudinal strength of hull. _____ are also longitudinal members situated at both sides of the _____ of bottom, which mainly withstand the total longitudinal bend. Floors play the part of chief transversal members at the bottom, which are subjected to _____ and gravity of cargoes & equipment _____ local external forces. They shoulder the responsibility of transversal and local strength of bottom.

V. Translation.

1. Translate the following sentences into Chinese.

(1) In addition to shell construction, there are bottom construction, side construction, deck construction, bulkhead construction, bow and stern construction and super construction, too.

(2) The composition of transversely framed single bottom construction is composed of bottom plating, center keelson, side keelson and floor, as shown in Fig.6.1.

(3) The longitudinal framed single bottom construction is composed of bottom plating, keelson, floor and a large number of bottom longitudinal, as shown in Fig.6.2.

(4) Freighters carrying general cargo cannot use this system since these transverse members would impede the stowage of cargo.

(5) As it were, the side constructions port and starboard are two side walls of hull symmetrical to each other, and among the side constructions leading components are the frames that bear water pressure and secure the transversal and local strength of hull.

2. Translate the short passage.

The double bottom is, in fact, watertight compartment. In the double bottom, the central stringer is a longitudinal member located at the centerline of bottom from bow to stern, which assures you of the total longitudinal strength of hull. Side girders are also longitudinal members situated at both sides of the centerline of bottom, which mainly withstand the total longitudinal bend. Floors play the part of chief transversal members at the bottom, which are subjected to water pressure and gravity of cargoes & equipment as well as local external forces. They shoulder the responsibility of transversal and local strength of bottom.

Module 4　Ship Drawings

Lesson 7　Deck and Bulkhead Construction

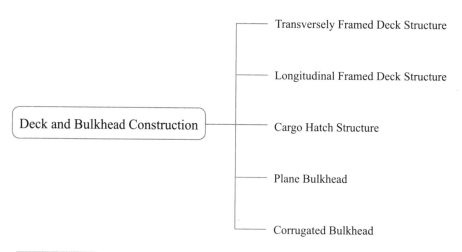

Background

A: Do you know the main hull of the ship in addition to the bottom and side?

B: In general, the ship's main hull has deck and bulkhead structures in addition to the bottom and side. Next we will study in further detail.

Text reading

1. Deck Structure

The deck structure is made up of deck plates, beams, deck girders, deck longitudinal, hatch, pillars and other members. Most of the deck structures are single planking structures, which can be divided into longitudinal framed deck structure and transversely framed deck structure according to the form of framing arrangement. There are cargo hatches, engine room hatches and other large openings and related buildings on the deck, so the structure is relatively complex. The continuous upper deck mainly bears the total longitudinal bending stress, so large ships generally adopt the longitudinal framed structure; the lower deck mainly bears the transverse loading, so most of them adopt the transversely framed structure.

（1）Transversely framed deck structure

The transverse strength of the transversely framed deck structure is good, and it is easy to manufacture. It is suitable for small ships, river ships and the lower deck of ships. The transversely framed deck structure is composed of deck plates, beams, deck girders, as shown in Fig.7.1.

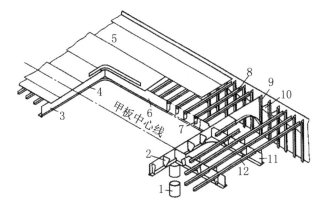

1—pillar; 2—tripping bracket; 3—hatch end beam; 4—steel rod; 5—deck; 6—hatch side girder; 7—bracket; 8—half beam; 9—main frame; 10—beam knee; 11—deck girder; 12—beam.

Fig. 7.1 Transversely Framed Deck Structure

（2）Longitudinal framed deck structure

The longitudinal strength of the longitudinal framed deck structure is good, but the assembly construction is more troublesome. It is mainly used for the upper deck of large and medium-sized ships with high requirements for the total longitudinal strength.

The longitudinal framed deck structure is composed of deck plates, deck longitudinal, deck girders and web beams. Fig.7.2 shows the longitudinal framed upper deck structure, in which the deck between hatches still adopts the transverse framed structure.

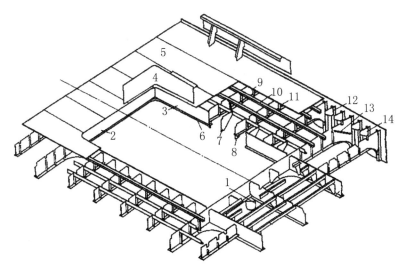

1—pillar; 2—hatch end beam; 3—hatch side girder; 4—hatch coaming; 5—upper deck; 6—steel rod; 7—tripping bracket; 8—small bracket; 9—web beam; 10—deck longitudinal; 11—stiffener rib; 12—frame; 13—diagonal stiffener; 14—bracket.

Fig. 7.2 Longitudinal Framed Deck Structure

（3）Cargo hatch structure

There is a large cargo hatch on the deck of the cargo ship. The hatch coaming is provided around the cargo hatch. Its function is to increase the strength of the hatch, prevent the sea water from pouring into the cabin, and ensure the safety of the operators.

2. Bulkhead Structure

There are many vertical division plates in the ship, which are called bulkheads. Next comes the bulkhead structure, which may assume the forms of either plain plate with stiffeners or corrugated plate. The former is composed of steel plates and stiffeners while the latter is steel plates pressed into corrugated shape. Bulkheads employed to prevent oil and water from leakage are put as oil-tight or water-tight bulkheads respectively. Bulkheads arranged along the ship length are called longitudinal bulkheads whereas those along the ship width are named transversal bulkheads.

The principal function of bulkheads is to separate the inner space of hull so as to guarantee vessel flood-ability and keep fire hazard and poisonous gas from spreading, not to mention the strength concern.

（1）Plane bulkhead

The plane bulkhead consists of bulkhead panels and frames, as shown in Fig. 7.3. The bulkhead frames are composed of two members: stiffeners and girders, which play a role in increasing the strength and rigidity of bulkhead panels.

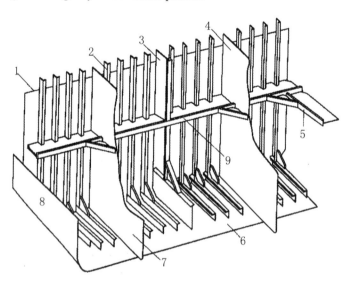

1—transverse bulkhead plating; 2—vertical stiffener; 3—vertical girder; 4—longitudinal bulkhead; 5—side stringer; 6—bottom plating; 7—longitudinal bulkhead; 8—side strake; 9—horizontal girder.

Fig. 7.3　Plane Bulkhead

（2）Corrugated bulkhead

The corrugated bulkhead is pressed by steel plate, and its groove-shaped bending is used instead of the role of stiffener (Fig.7.4).

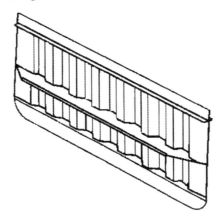

Fig. 7.4　corrugated bulkhead

Compared with the plane bulkhead, the characteristics of the corrugated bulkhead are: under the condition of ensuring the same strength, the corrugated bulkhead structure of small and medium-sized ships is light in weight and steel is saved; there are fewer parts forming the corrugated bulkhead, which can reduce the workload of assembly and welding; it is convenient for cleaning the tank and is conducive to preventing corrosion. However, the pressure-bearing capacity perpendicular to the corrugated direction is relatively poor, and the corrugated bulkhead occupies a large volume, which is not conducive to loading miscellaneous cargo. Therefore, the corrugated bulkhead is suitable for oil tankers, bulk carriers, as well as for container ships and general cargo ships when the tank depth is large.

New Words and Expressions

1. deck girders 甲板纵桁
2. pillar ['pɪlə(r)] n. 支柱
3. longitudinal framed deck structure 纵骨架式甲板结构
4. manufacture [ˈmænjuˈfæktʃə(r)] vt. 制造
5. tripping bracket 防倾肘板
6. hatch end beam 舱口端横梁
7. steel rod 圆钢
8. beam knee 梁肘板
9. hatch side girder 舱口纵桁
10. hatch coaming 舱口围板
11. web beam 强横梁
12. stiffener rib 加强筋
13. diagonal stiffener 斜置加强筋
14. vertical stiffener 垂直扶强材

15. side stringer 舷侧纵桁
16. corrugated bulkhead 槽形舱壁
17. assembly [əˈsembli] *n.* 装配
18. conducive [kənˈdjuːsɪv] *adj.* 有益的
19. corrosion [kəˈrəʊʒ(ə)n] *n.* 腐蚀
20. miscellaneous [ˌmɪsəˈleɪnɪəs] *adj.* 多方面的；各种的

Notes

1. Most of the deck structures are single planking structures, which can be divided into longitudinal framed deck structure and transversely framed deck structure according to the form of framing arrangement.
   译：甲板大部分是单层板架结构，按骨架设置形式可分为纵骨架式和横骨架式甲板结构。

2. The transverse strength of the transversely framed deck structure is good, and it is easy to manufacture. It is suitable for small ships, river ships and the lower deck of ships.
   译：横骨架式甲板结构的横向强度好，制造方便，适用于小型船舶、内河船及船舶的下甲板。

3. Bulkheads employed to prevent oil and water from leakage are put as oil-tight or water-tight bulkheads respectively.
   译：用来防止油和水泄漏的舱壁分别采用油密或水密舱壁设置。

Expanding reading

Fore Ship

In respect to the bow structure and stern structure, the former usually denotes the zone from bow up to 0.15 times of the ship length starting from fore perpendicular to aft, while the latter indicates the zone after the bulkhead of aft peak.

The fore ship (Fig.7.5) is the part of the ship between the stem and the collision or forepeak bulkhead. The space in front of the collision bulkhead is the forepeak. The forepeak tank is usually used as a ballast tank. If the ship is not loaded, this is often filled with water to reduce the trim at the stern. Often there is a wash bulkhead in the peak tanks. This improves the rolling behaviour of the ship, by delaying movement of the ballast water when the tanks are not completely filled. on the top of the forepeak, right below the capstan or anchor winch there are chain lockers for the storage of the anchor chains. Above the weather deck on the foreship is the forecastle. On the forecastle is the windlass and other mooring equipment, also the foremast.

Lesson 7 Deck and Bulkhead Construction

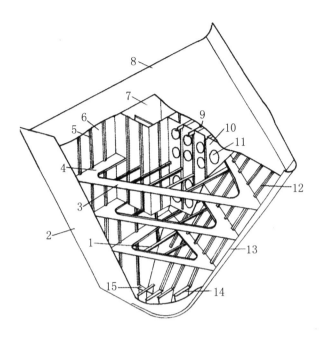

1—panting beam; 2—side plate; 3—side stringer; 4—horizontal girder; 5—stiffener; 6—forepeak bulkhead; 7—chain locker; 8—deck; 9—beam; 10—centerline wash bulkhead; 11—lightening hole; 12—frame; 13—stem post; 14—rising floor; 15—intercostal center keelson.

Fig. 7.5 Fore Ship

The foreship is subject to extra large forces and stress that are caused by:

A. The pitching of the ship;

B. The foreship moving in and out of the water (panting stresses);

C. Maintaining speed in heavy weather;

D. Ice.

To compensate for these forces, the foreship needs additional reinforcements that sometimes partly extend to the back of the ship. The bulb stem is added to reduce the wave resistance. The bow wave causes a resistance that has a negative influence on the speed of the ship. The additional bulb creates a wave, which extensively equalizes the bow wave and thus positively influences the ship's speed (only under the loaded condition). It is most effective at a certain draught (loaded ship), but not so ideal in the case of an unloaded ship because the bulb will actually produce more resistance.

New Words and Expressions

1. collision bulkhead 防撞舱壁
2. capstan[ˈkæpstən] *n.* 起锚机，绞盘
3. anchor winch 锚机，绞车，起锚绞车
4. chain locker 锚链舱
5. center line wash bulkhead 中纵制荡舱壁

6. stem post 首柱

Exercises

Ⅰ. Answer the following questions according to the passage.

1. What could the ship deck be divided into by the direction of framing arrangement?
_____

2. What is the advantage of transversely framed deck structure?
_____

Ⅱ. Practic these new words.

1. English to Chinese.

   deck girders _____   hatch end beam _____

   steel rod _____   tripping bracket _____

   beam knee_____   hatch coaming _____

2. Chinese to English.

   斜置加强筋 _____   舱口纵桁 _____

   垂直扶强材 _____   槽型舱壁 _____

   腐蚀 _____   水平扶强材 _____

Ⅲ. Explain the nouns.

1. hatch end beam
_____

2. beam knee
_____

Ⅳ. Fill in the blanks with the proper words or expressions given below.

> assembly, forming, perpendicular, ensuring,
> corrugated bulkhead, bulk carriers, conducive

Compared with the plane bulkhead, the characteristics of the _____ are: under the condition of _____ the same strength, the corrugated bulkhead structure of small and medium-sized ships is light in weight and steel is saved; there are fewer parts _____ the corrugated bulkhead, which can reduce the workload of _____ and welding; it is convenient for cleaning the tank and is conducive to preventing corrosion. However, the pressure-bearing capacity _____ to the corrugated direction is relatively poor, and the corrugated bulkhead occupies a large volume, which is not _____ to loading miscellaneous cargo. Therefore, the corrugated bulkhead is suitable for oil tankers, _____, as well as for container ships and general cargo ships when the tank depth is large.

V. Translation.

1. Translate the following sentences into Chinese.

(1) The longitudinal strength of the longitudinal framed deck structure is good, but the assembly construction is more troublesome.

(2) Its function is to increase the strength of the hatch, prevent the sea water from pouring into the cabin, and ensure the safety of the operators.

(3) Next comes the bulkhead structure, which may assume the forms of either plain plate with stiffeners or corrugated plate.

(4) The principal function of bulkheads is to separate the inner space of hull so as to guarantee vessel flood-ability and keep fire hazard and poisonous gas from spreading, not to mention the strength concern.

(5) The bulkhead frames are composed of two members: stiffeners and girders, which play a role in increasing the strength and rigidity of bulkhead panels.

2. Translate the short passage.

The deck structure is made up of deck plates, beams, deck girders, deck longitudinal, hatch, pillars and other members. Most of the deck structures are single planking structures, which can be divided into longitudinal framed deck structure and transversely framed deck structure according to the form of framing arrangement. There are cargo hatches, engine room hatches and other large openings and related buildings on the deck, so the structure is relatively complex. The continuous upper deck mainly bears the total longitudinal bending stress, so large ships generally adopt the longitudinal framed structure; the lower deck mainly bears the transverse loading, so most of them adopt the transversely framed structure.

Lesson 8 Lines Plan

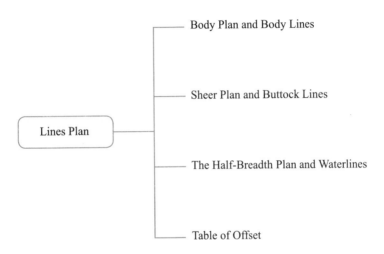

Background

A: Do you know how to describe the exact geometrical shape of the ship?

B: Usually, the shape of a ship is represented graphically by a lines drawing which consists of projections of the intersection of the hull with a series of planes. Next we will study in further detail.

Text reading

During the conceptual design and preliminary design, when the principal dimensions, displacement and form coefficients are known, one will have an impressive amount of design information, but not yet a clear image of the exact geometrical shape of the ship. Because of the complexity, a ship's hull cannot be written that fully describes the shape of a ship. Usually, the shape of a ship is represented graphically by a lines drawing which consists of projections of the intersection of the hull with a series of planes. The planes are equally spaced in each of the three dimensions (length, breadth and height). Planes in one dimension will be perpendicular to planes in the other two dimensions. In other words, the sets of planes are mutually perpendicular or orthogonal planes.

Nowadays the lines plans are being made with the aid of computer programs that possess possibility to transform the shape of a vessel automatically when modifications in the ship's design require this. When a lines plan is ready, the programs may be used to calculate, among other things, the volume and stability of the ship (Fig.8.1).

Fig. 8.1　The imaginary box

1. Body Plan and Body Lines

Planes parallel to the front and back of the imaginary box running port to starboard are called stations. A ship is typically divided into 11 or 21 evenly spaced stations. The larger the ship, the more stations will be made. The afermost station is called the after perpendicular(AP) and labeled station number zero(ordinate 0). The first forward station at the bow is called the forward perpendicular (FP , ordinate 20). The station midway between the perpendiculars is called the midships station, usually represented by the ⊗ symbol. Each station plane will intersect the ship's hull and form a curved line at the points of intersection. These lines are called ordinate lines. A projection of all ordinates into one view is called a body plan. Usually additional half ordinates are also drawn at the ends where a greater change of shape occurs. A half transverse section only is drawn since the vessel is symmetrical about the centre line , and forward half sections are drawn to the right of the centre line with aft half sections to the left.

Each body line shows the true shape of the hull from the aft view for some longitudinal position on the ship which allows this line to serve as a pattern for the construction of the ship's transverse framing. The projection of a set of transverse lines reflects the variation of the shape of the outer plate surface along the captain's direction. The grid network on the body plan is straight lines that represent the orthogonal planes containing the buttock lines and waterlines (Fig.8.2).

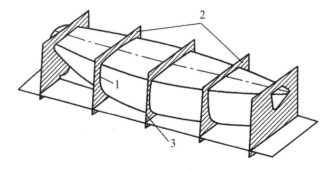

1—Body Lines; 2—transverse plane; 3—midstation plane.

Fig. 8.2　Body Plan and Body Lines

2. Sheer Plan and Buttock Lines

A plane that runs from bow to stern directly through the center of the ship and parallel to the sides o f the imaginary box is called the center line plane. A series of planes parallel to one side of the center line plane are imagined at regular intervals from the center line. Each plane will intersect the ship's hull and form a curved line at the points of intersection.These lines are called buttock line and are all projected onto a single plane called the sheer Plan.

Each buttock line shows the true shape of the hull from the right side view for some distance from the center line of the ship. This allows them to serve as a pattern for the construction of the ship's longitudinal framing. The grid network on the sheer plan is straight line that represent the orthogonal planes containing the station lines and waterlines (Fig. 8. 3).

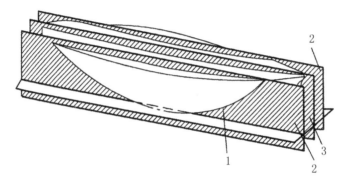

1—Buttock Lines; 2—longitudinal plane; 3—center-line plane

Fig. 8.3 Sheer Plan and Buttock Lines

3. The Half-Breadth Plan and Waterlines

The bottom of the box is a reference plane called the base plane. The base plane is usually level with the keel. A series of planes parallel and above the base plane are imagined at regular intervals. Each plane will intersect the ship's hull and form a line at the points of intersection. These lines are called waterlines and are all projected onto a single plane called the half breadth plan, since ships are symmetric about their center-line they only need be drawn for the port(usually) or starboard side. There will be one plane above the base plane that coincides with the normal draft (draught) of the ship, this waterline is called the designed waterline or loaded waterline. This waterline is actually the waterline the ship floats. The designed water line is often represented on drawings as DWL. (Fig.8.4)

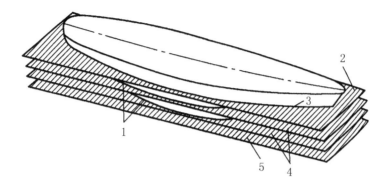

1—Waterlines; 2—designed waterplane; 3—designed waterline; 4—horizontal plane; 5—base plane.

Fig. 8.4 The Half-Breadth Plan and Waterlines

Each waterline shows the true shape of the hull from the top view for some elevation above the base plane which allows this line to serve as a pattern for the construction of the ship's framing. The grid network on the half-breadth plan is straight lines that represent the orthogonal planes containing the buttock and station lines.

4. Table of Offset

When the lines plan was completed the draftsmen would compile a "table of offsets", which converted from the information in the drawing to a numerical representation, for calculating geometric characteristics of the hull such as sectional area, water-plane area, submerged volume and the longitudinal center of flotation (Fig.8.5).

stations	The half-breadth value of the central longitudinal section (Y)									
	1,100 Waterline	2,200 Waterline	3,300 Waterline	4,400 Waterline	Designed waterline	6,600 Waterline	upper deck side line	Fore castle deck side line	Poop deck side line	bulwark top line
0	—	—	—	—	1,097	2,233	3,806	—	—	3,896
1	491	662	978	1619	2,602	3,498	4,689	—	5,621	562
2	1,372	1,882	2,501	3,271	4,003	4,635	5,444	—	6,273	6,273
3	2,595	3,455	4,152	4,732	5,181	5,583	6,066	—	6,549	6,549
4	4,062	5,001	5,510	5,817	6,066	6,288	6,517	—	6,698	6,698
5	5,308	6,017	6,304	6,470	6,588	6,690	6,787	—	6,841	6,841
6	6,075	6,537	6,717	6,794	6,831	6,858	6,881	—	—	6,900
7	6,521	6,799	6,884	6,899	6,900	6,900	6,900	—	—	6,900
8	6,732	6,891	6,900	6,900	6,900	6,900	6,900	—	—	6,900
9	6,780	6,900	6,900	6,900	6,900	6,900	6,900	—	—	6,900
10	6,780	6,900	6,900	6,900	6,900	6,900	6,900	—	—	6,900
11	6,780	6,900	6,900	6,900	6,900	6,900	6,900	—	—	6,900
12	6,720	6,896	6,900	6,900	6,900	6,900	6,900	—	—	6,900
13	6,508	6,780	6,855	6,882	6,896	6,900	6,900	—	—	6,900
14	6,108	6,464	6,599	6,694	6,763	6,822	6,870	—	—	6,899
15	5,390	5,861	6,073	6,220	6,335	6,444	6,598	—	—	6,735
16	4,222	4,857	5,127	5,304	5,449	5,621	5,994	—	—	6,300
17	2,805	3,504	3,853	4,049	4,234	4,491	5,096	—	—	5,578
18	1,472	2,105	2,452	2,668	2,862	3,510	4,011	5,234	—	5,301
19	378	800	1,076	1,271	1,145	689	2,755	4,080	—	—
20	—	—	—	—	—	322	1,380	2,710	—	3,636

Fig. 8.5 Table of Offset

The offset data is the distance measured from the center line of the ship to the specific point on the curves (station or waterline curves). The offset data must be measured at every intersection points on each stations and waterlines. Offset data also called as half breadth data, because it represents the half-breadth of the ship at every station and waterlines.

Since the original lines plan was drawn in a small scale which varied with the size of ship, the amplified mistakes and errors become obvious after lofting which makes it necessary to modify the lines plan and offsets. Then the offsets used for building the ship would subsequently be lifted by the lofts-man from the full size or 10 to 1 scale lines for each frame.

New Words and Expressions

1. conceptual [kənˈseptʃʊəl] *adj.* 观念的，概念的
2. preliminary [prɪˈlɪmɪnəri] *adj.* 初步，初始
3. geometrical [dʒɪəˈmetrɪkl] *adj.* 几何的，几何学的
4. represent [ˌreprɪˈzent] *v.* 代表；表现；描写
5. intersection [ˌɪntəˈsekʃn] *n.* 交集；十字路口；交叉点
6. perpendicular [ˈpɜːpənˈdɪkjələ(r)] *adj.* 垂直的；直立的；陡峭的
7. orthogonal [ɔrˈθɑɡənəl] *adj.* 直角的；矩形的；直交的；互相垂直的
8. body plan and body lines 横剖线图和横剖线
9. symmetrical [sɪˈmetrɪk(ə)l] *adj.* 匀称的，对称的
10. sheer plan and buttock lines 纵剖线图和纵剖线
11. parallel [ˈpærəlel] *adj.* 平行的；相同的，类似的
12. interval [ˈɪntəvl] *n.* 间隔；（数学）区间
13. the half-breadth plan and waterline 半宽水线图和水线
14. keel [kiːl] *n.* 龙骨；平底船；龙骨脊
15. coincide [ˌkəʊɪnˈsaɪd] *vi.* 与……一致；想法、意见等相同；相符；极为类似
16. elevation [ˌelɪˈveɪʃ(ə)n] *n.* 提升；海拔；高度；提高；立视图
17. table of offset 型值表
18. draftsmen [ˈdrɑːftsmən] *n.* 绘图员
19. sectional area 横剖面面积
20. water plane area 水线面面积
21. submerged volume 水下体积
22. the longitudinal center of flotation 浮心纵向位置

Notes

1. During the conceptual design and preliminary design, when the principal dimensions, displacement and form coefficients are known, one will have an impressive amount of design information, but not yet a clear image of the exact geometrical shape of the ship.

译：在概念设计和初步设计阶段，当主尺寸、排水量和形状系数已知时，人们会获得大量的设计信息，但还不能清楚地描绘出船的确切几何形状。

2. The projection of a set of transverse lines reflects the variation of the shape of the outer plate surface along the captain's direction. The grid network on the body plan is straight lines that represent the orthogonal planes containing the buttock lines and waterlines.

译：一组横剖线的投影反映了外板型表面形状沿船长方向的变化情况。横剖线图上的格子线是直线，表示包含纵剖线和水线的正交平面。

3. When the lines plan was completed the draftsmen would compile a "table of offsets", which converted from the information in the drawing to a numerical representation, for calculating geometric characteristics of the hull such as sectional area , water-plane area, submerged volume and the longitudinal center of flotation.

译：当型线图绘制完成后，绘图员编制"型值表"，将图中的信息转化为数字表示，计算船体的横剖面面积、水线面面积、水下体积和浮心纵向位置等几何特征。

Expanding reading

Frame Body Plan

Companied with the shell expansion plan, the frame body plan shows layout of plates and location of major members. The molded shape of the ship, seams, decks, platforms and the arrangement of longitudinal members are included in this drawing which serves as a valuable base and reference for other drawings and hull construction.

Compared with body plan, this plan is obviously different. In body plan, ordinate lines represent the intersection lines of the 10 or 20 evenly divided transverse sections and the hull surface. However, the frame body plan, based on the lines plan, displays the intersected lines where each frame located. In other words , the true shape of each frame line is shown here. Same as body plan, the aft frames are drawn to the left part while the forward to the right because of the symmetry of the hull. Typically, only odd frames are drawn and every frame at the end is necessary since the sharp change at the stem and stern.

In addition to the horizontal lines (water lines) and vertical lines (buttock lines) , curves can be categorized as frame lines, seams of shell plates, intersection lines of members and imaginary lines (e. g. the top line of bilge brackets).

The plate seams show the arrangement and projection shape of shell plates. Usually there are three kinds: seams, butts, block seams.

Lesson 8 Lines Plan

New Words and Expressions

1. frame body plan 肋骨型线图
2. hull surface 船体表面
3. symmetry [ˈsɪmətri] *n.* 对称
4. odd [ɒd] *adj.* 奇数
5. buttock line 纵剖线
6. imaginary line 假想线
7. bilge bracket 舭肘板
8. projection [prəˈdʒekʃn] *n.* 投影
9. block seam 分段接缝

Exercises

Ⅰ. Answer the following questions according to the passage.

1. What is the function of lines plan?
   _____

2. What kind of information can you get from the lines plan?
   _____

Ⅱ. Practic these new words.

1. English to Chinese.

   table of offset _____   draftsmen _____

   sectional area _____   water plane area _____

   submerged volume _____   intersection _____

2. Chinese to English.

   半宽水线图 _____   水线 _____

   纵剖线图 _____   纵剖线 _____

   横剖线图 _____   横剖线 _____

Ⅲ. Explain the nouns.

1. table of offset

_____

2. the half-breadth plan

_____

Ⅳ. Fill in the blanks with the proper words or expressions given below.

> intersection, keel, parallel, the half breadth plan,
> designed waterline, the base plane

The bottom of the box is a reference plane called _____. The base plane is usually level with the _____. A series of planes _____ and above the base plane are imagined at regular intervals. Each plane will intersect the ship's hull and form a line at the points of _____.These lines are called waterlines and are all projected onto a single plane called _____, since ships are symmetric about their center line they only need be drawn for the port (usually) or starboard side. There will be one plane above the base plane that coincides with the normal draft (draught) of the ship, this waterline is called the _____ or loaded waterline. This waterline is actually the waterline the ship floats. The designed water line is often represented on drawings as DWL.

Ⅴ. Translation.

1. Translate the following sentences into Chinese.

(1) Usually, the shape of a ship is represented graphically by a lines drawing which consists of projections of the intersection of the hull with a series of planes.

_____

(2) When a lines plan is ready, the programs may be used to calculate, among other things, the volume and stability of the ship.

_____

(3) A series of planes parallel to one side of the center line plane are imagined at regular intervals from the center line.

_____

(4) Each waterline shows the true shape of the hull from the top view for some elevation above the base plane which allows this line to serve as a pattern for the construction of the ship's framing.

(5) The offset data is the distance measured from the center line of the ship to the specific point on the curves (station or waterline curves).

2. Translate the short passage.

Planes parallel to the front and back of the imaginary box running port to starboard are called stations. A ship is typically divided into 11 or 21 evenly spaced stations. The larger the ship, the more stations will be made. The afermost station is called the after perpendicular (AP) and labeled station number zero(ordinate 0). The first forward station at the bow is called the forward perpendicular (FP , ordinate 20). The station midway between the perpendiculars is called the midships station, usually represented by the ⊗ symbol. Each station plane will intersect the ship's hull and form a curved line at the points of intersection. These lines are called ordinate lines. A projection of all ordinates into one view is called a body plan. Usually additional half ordinates are also drawn at the ends where a greater change of shape occurs. A half transverse section only is drawn since the vessel is symmetrical about the centre line , and forward half sections are drawn to the right of the centre line with aft half sections to the left.

Lesson 9　General Arrangement Plan

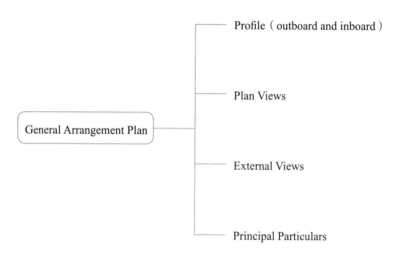

Background

A: Do you know the general arrangement plan?
B: Usually, you can learn the form of the superstructure and the arrangement of the cabins, equipment, doors, windows, passages, etc. Next we will study in further detail.

Text reading

　　The general arrangement plan is a kind of general arrangement of the whole ship, which reflects the technical and economic performance of ship and is one of the important basic drawings of the whole ship. From the general arrangement plan, you can learn the form of the superstructure and the arrangement of the cabins, equipment, doors, windows, passages, etc. When calculating the weight and position of the center of gravity of the whole ship, designing the equipment and structure of the ship, the general arrangement plan is the basis of designing and calculation. The general arrangement plan is also the basis for drawing other drawings, such as all kinds of equipment, system arrangement, doors, windows, escalator arrangement, wood and insulation arrangement, etc.

　　The general arrangement plan can be used as a guiding drawing in the concrete construction. It can play the coordination of the machinery, equipment, when they have contradictions, the general arrangement plan is referred. In addition, the general arrangement plan is of great significance to the outfitting work of ship construction. Therefore, it is necessary to correctly read and draw the general arrangement.

The efficient operation of a ship depends upon the proper arrangement of each separate space and the most effective interrelationships among all spaces. The general arrangement plan, or GA as it is commonly called, depicts the assignment of spaces for all the required functions and equipment, properly coordinated for location and access. It is important that the general arrangement plan be functionally and economically developed with respect to factors that affect both the construction and operation cost, especially the manpower required to operate the ship.

The GA consists of (at a minimum) a plan view of each major deck of the vessel, shows all of the watertight and structural bulkheads, as well as joiner bulkheads. All of the furniture is typically shown or in early stages the furniture and large items to be on the vessel are roughly blocked in (though this may be broke out into arrangement drawings for complex arrangements). Passageways, stairwells and all equipment vital to the ships operation are shown.

1. Profile (outboard and inboard)

The profile, a side-view of the ship, is important in conveying the general geometry of deckhouse & superstructures, the indication of decks and equipment heights (Fig.9.1). It also shows a cut-away view through the centre line of the vessel, inclusive of centre line structures. The profile is a front view from the starboard side of the ship and is usually drawn at the top of the drawing. The profile is the main view of the general arrangement plan, and the basic content it expresses is:

(1) It shows the appearance of the side of the ship.

(2) It expressed the outline of the main compartment division.

(3) It expressed an overview of the ship equipment layout.

(4) It expressed the layout of doors, windows and escalators.

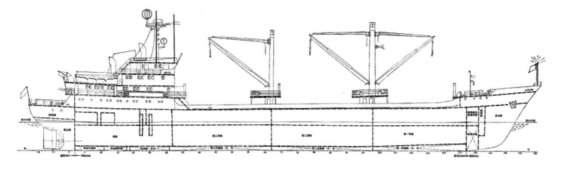

Fig. 9.1 Profile

2. Plan Views

The GA has, at a minimum, plan views of the most important decks. They are usually drawn below the profile and arranged from top to bottom according to the position of deck, platform and bilge (Fig.9.2).

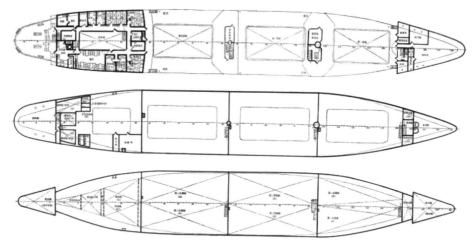

Fig.9.2 Profile Plan Views

The basic content of the plan views is:

(1) On the deck or platform, the division of cabins, the arrangement of cabin equipment, appliances, etc., and the positions of these cabins and equipment, appliances, etc. in the direction of the length and width of the ship.

(2) The arrangement of ship equipment and machinery on the deck or platform, outside the cabin, and the position of these equipment and machinery in the direction of the length and width of the ship.

(3) Arrangement of passages, doors, windows, escalators, etc. on decks or platforms.

3. External Views

Sometimes a front and back view are included. End views are often added, particularly where a vessel has bow or stern operated equipment. Top views are occasionally seen, particularly on vessels with complex towing arrangements, or extensive exterior deck spaces. Bottom views are rarely used unless the vessel has a complex or unique propulsion arrangement.

4. Principal Particulars

Additionally, some basic data are also included in the drawing like: principal dimensions, volumes of the holds, tonnage, displacement, dead weight, engine power, speed and class. Fig.9.3 shows an example of some important information: length OA, beam MLD, depth MLD, draft designed, complement.

Lesson 9　General Arrangement Plan

主尺度

PRINCIPAL DIMENSIONS

总长 L_{OA}	约 84.25 m
L_{OA}	abt. 84.25 m
垂线间长 L_{bp}	76.80 m
L_{bp}	76.80 m
型宽 B	14.50 m
BREADTH(MLD.)	14.50 m
型深 D	9.30 m
DEPTH(MLD)	9.30 m
设计吃水 T_d	6.10 m
DRAFT(DESIGN)	6.10 m
结构吃水 T_s	6.40 m
DRAFT(SCANTLING)	6.40 m
载重量 DWT	约 4 000 t
DEADWEIGHT	abt. 4 000 t
乘员	18 人
COMPLEMENT	18P

Fig.9.3　Principal Particulars

New Words and Expressions

1. arrangement [əˈreɪndʒmənt] *n.* 安排；排列；约定；布置
2. cabin [ˈkæbɪn] *n.* 小木屋；客舱；（轮船上工作或生活的）隔间
3. passage [ˈpæsɪdʒ] *n.* （文章的）一段；经过；通路，通道；旅程，行程
4. escalator [ˈeskəleɪtə(r)] *n.* 自动扶梯
5. insulation [ˌɪnsəˈleɪʃn] *n.* 隔离，孤立；绝缘；绝缘或隔热的材料；隔声
6. contradiction [ˌkɒntrəˈdɪkʃn] *n.* 矛盾；反驳；否认；不一致
7. joiner bulkhead 轻型甲板
8. profile [ˈprəʊfaɪl] *n.* 侧面；轮廓；外形；剖面；侧面图
9. plan views 平面图
10. bilge [bɪldʒ] *n.* 舱底
11. external views 外部视图
12. end views 端视图
13. top views 俯视图
14. occasionally [əˈkeɪʒnəli] *adv.* 偶尔；间或
15. towing arrangement 拖曳设备

16. exterior [ɪkˈstɪəriə(r)] *adj.* 外部的；表面的；外在的

17. propulsion arrangement 推进设备

18. complement [ˈkɑːplɪment] *n.* 补充；船员

Notes

1. The general arrangement plan is a kind of general arrangement of the whole ship, which reflects the technical and economic performance of ship and is one of the important basic drawings of the whole ship.

   译：船体总布置图是表示全船总体布置的图样，它比较集中地反映了船舶的技术、经济性能，是重要的全船性基本图样之一。

2. In addition, the general arrangement plan is of great significance to the outfitting work of ship construction. Therefore, it is necessary to correctly read and draw the general arrangement.

   译：另外，总布置图对船舶建造时的舾装工作尤其有着重要意义。

3. The GA consists of (at a minimum) a plan view of each major deck of the vessel, shows all of the watertight and structural bulkheads, as well as joiner bulkheads.

   译：总布置图（至少）包括船舶各主要甲板的平面图，显示所有水密舱壁、结构舱壁和轻型舱壁。

Expanding reading

Shell Expansion Plan

Shell expansion plan is a two-dimensional drawing showing the arrangement of the shell plates, stiffening members, seams. The purpose of the drawing is to assist in the plate development and subsequently the cutting of the strakes prior to fabrication. Because the shell is expanded only in the transverse direction, all vertical dimensions in this drawing are taken around the girth of the vessel rather than their being a direct vertical projection. This technique illustrates both the side and bottom plating as a continuous whole. The longitudinal curvature of the hull is small enough to be ignored which means the projection length in the plan represents (not equal to) the actual length. The transverse expanded length is equal to the real length of the shell, i.e., the width of each plate in the plan is the actual width.

In Fig.9.4 a typical shell expansion for a tanker is illustrated. This also shows the numbering of plates, and lettering of plate strakes for reference purposes and illustrates the system where strakes "run out" as the girth decreases forward and aft. This drawing was often subsequently retained by the ship owner to identify plates damaged in service. Since prefabrication became the accepted practice any shell expansion drawing produced will generally have a numbering system related to the erection of fabrication units rather than individual plates. However single plates were often marked in sequence to aid ordering and production identification.

Lesson 9 General Arrangement Plan / 75

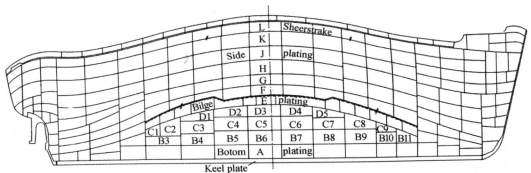

FRAMING, STRINGERS, DECKS AND OPENINGS IN SIDE SHELL ARE ALSO SHOWN ON THE SHELL EXPANSION BUT HAVE BEEN OMITTED FOR CLARITY

Fig. 9.4 Shell expansion plan

New Words and Expressions

1. shell expansion plan 外板展开图
2. two-dimensional 二维
3. subsequently [ˈsʌbsɪkwəntli] adv. 之后，接着
4. fabrication [ˌfæbrɪˈkeɪʃn] n. 制造，建造，装配
5. girth [gɜːrθ] n. 围长
6. illustrate [ˈɪləstreɪt] v. 表明
7. curvature [ˈkɜːvətʃə(r)] n. 弯曲，曲度
8. identify [aɪˈdentɪfaɪ] v. 确认，鉴定
9. prefabrication [ˌpriːfæbrɪˈkeɪʃən] n. 预装配
10. sequence [ˈsiːkwəns] n. 顺序，序列

Exercises

I. Answer the following questions according to the passage.

1. What is the function of the general arrangement plan?

2. What kind of information can you get from the general arrangement plan?

II. Practic these new words.

1. English to Chinese.

plan views _____   external view _____

end views _____   top view _____

propulsion arrangement_____   joiner bulkhead _____

2. Chinese to English.

总布置图 _____   上层建筑 _____

侧面图 _____   右舷 _____

艏部 _____   艉部 _____

III. Explain the nouns.

1. the general arrangement plan.

   _____

2. profile.

   _____

IV. Fill in the blanks with the proper words or expressions given below.

> **passages, insulation, superstructure, cabins, escalator, arrangement, contradictions**

The general _____ plan is a kind of general arrangement of the whole ship, which reflects the technical and economic performance of ship and is one of the important basic drawings of the whole ship. From the general arrangement plan, you can learn the form of the _____ and the arrangement of the _____, equipment, doors, windows, _____, etc. When calculating the weight and position of the center of gravity of the whole ship, designing the equipment and structure of the ship, the general arrangement plan is the basis of designing and calculation. The general arrangement plan is also the basis for drawing other drawings, such as all kinds of equipment, system arrangement, doors, windows, _____ arrangement, wood and _____ arrangement, etc.

The general arrangement plan can be used as a guiding drawing in the concrete construction. It can play the coordination of the machinery, equipment, when they have _____, the general arrangement plan is referred. In addition, the general arrangement plan is of great significance to the outfitting work of ship construction. Therefore, it is necessary to correctly read and draw the general arrangement.

V. Translation.

1. Translate the following sentences into Chinese.

(1) The general arrangement plan, or GA as it is commonly called, depicts the assignment of spaces for all the required functions and equipment, properly coordinated for location and access.

_____

(2) The profile, a side-view of the ship, is important in conveying the general geometry of deckhouse & superstructures, the indication of decks and equipment heights.

_____

(3) The profile is a front view from the starboard side of the ship and is usually drawn at the top of the drawing.

_____

(4) The GA has, at a minimum, plan views of the most important decks. They are usually drawn below the profile and arranged from top to bottom according to the position of deck, platform and bilge.

_____

(5) Sometimes a front and back view are included. End views are often added, particularly where a vessel has bow or stern operated equipment.

_____

2. Translate the short passage.

Profile (outboard and inboard)

　　The profile, a side-view of the ship, is important in conveying the general geometry of deckhouse & superstructures, the indication of decks and equipment heights. It also shows a cut-away view through the centre line of the vessel, inclusive of centre line structures. The profile is a front view from the starboard side of the ship and is usually drawn at the top of the drawing. The profile is the main view of the general arrangement plan, and the basic content it expresses is:

　　(1) It shows the appearance of the side of the ship.
　　(2) It expressed the outline of the main compartment division.
　　(3) It expressed an overview of the ship equipment layout.
　　(4) It expressed the layout of doors, windows and escalators.

_____

_____

Lesson 10　Basic Structural Plan

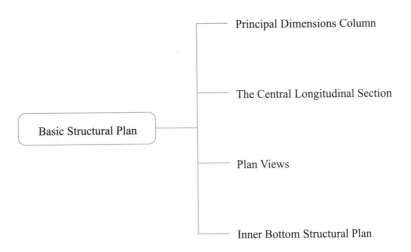

Background

A: What do you know about the basic structure plan?
B: Usually, the basic structure plan is a whole ship-like structural drawing, which is the basis for drawing other structural drawings and is also an instructional drawing during construction. Next we will study in further detail.

Text reading

The basic structure plan is composed of a longitudinal section view and several horizontal section views or sectional views, it is the basic drawings showing the structure of the hull. It forms a three-dimensional view showing the structure of the whole ship with the midship section plan, thereby completely expressing the size and structural form of the main longitudinal and transverse members of the whole ship. The basic structure plan is a whole ship-like structural drawing, which is the basis for drawing other structural drawings and is also an instructional drawing during construction. Therefore, the basic structure plan of the hull must be correctly read and drawn. The basic structure plan consists of principal dimensions column and a set of views.

1. Principal Dimensions Column

The main contents of principal dimensions column in the basic structure plan are: length over all (L_{OA}), length on the water line (L_{wl}), length between perpendiculars (L_{pp}), breadth, depth, draft, frame spacing, and height between decks. It should be noted at the top right of the drawing.

2. The Central Longitudinal Section

The longitudinal sectional view of the basic structural plan is a longitudinal sectional view obtained by cutting the hull through a plane parallel to the hull center plane or a plane near the center plane. The cross-sectional view obtained by cutting the hull through the center line plane of the hull, also known as the central longitudinal section, usually adopts this form. The longitudinal section view is generally arranged above the drawing surface, showing the arrangement of the hull members in the length and depth of the ship and the connection of some longitudinal members. The central longitudinal section refers to a side view, i.e. profile along the centre line. Lots of information can be found here (Fig.10.1):

(1) The structural types, sizes and connection ways of members in center plane. The projection profiles of these members are expressed by thin solid or dashed lines.

(2) Members between the center plane and the side of ship are overlap projected, using thin double dots dash lines for visible outlines and thin dashed lines for invisible outlines.

(3) Members located at the side of ship can be simplified as thick dot dash lines for side girders and web frames, thin dot dash lines for tween frames. The general frames can be omitted.

(4) Members across the center line usually are cut here by using thick solid lines.

Fig. 10.1　The Central Longitudinal Section

3. Plan Views

The plan views provide a series of planes of decks or platforms which obtained nearly above the surfaces of themselves.

It is a cross-sectional view obtained by cutting the hull along the upper surface of the deck or platform with a cut plane, as shown in section A-A ~ I-I in Fig.10.2. Deck and platform diagrams are a set of views, arranged in the middle of the drawing, which represent the structure of the deck and platform and the components directly connected to it.

(1) The arrangements and thicknesses of deck (platform) plates, the positions and sizes of stiffened panels and openings. Thin solid lines are used for seams and outlines of openings, while thin solid lines with oblique lines for stiffened panels.

(2) Members on decks (or platforms).

(3) The longitudinal and transverse bulkhead, coaming and stiffened beam knees. Usually, thick solid lines are used for visible bulkheads and coaming, rail lines for invisible watertight members, and thick dashed lines for invisible non-watertight members and knees.

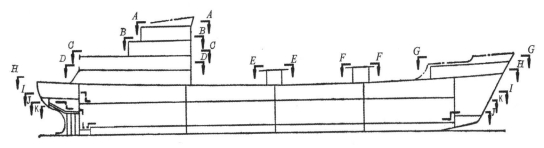

Fig. 10.2   Selection of platforms, decks and bottom sections

4. Inner Bottom Structural Plan

There are usually two forms of expression of the inner bottom structural plan (Fig.10.3): one is a cross-sectional view obtained by cutting the hull from a cut plane between the lowest deck and the bottom frame, so that the side structures that must be expressed can be clearly expressed, such as Fig.10.2 K-K section; the second is the method of stepwise sectioning by cutting the hull from a plan view near the upper edge of the bottom frame (the double bottom part uses a cutting plane near the upper surface of the inner bottom plate, and the single bottom part uses a cutting plane near the floor top) This view is relatively simple, such as J-J in Fig.10. 2.

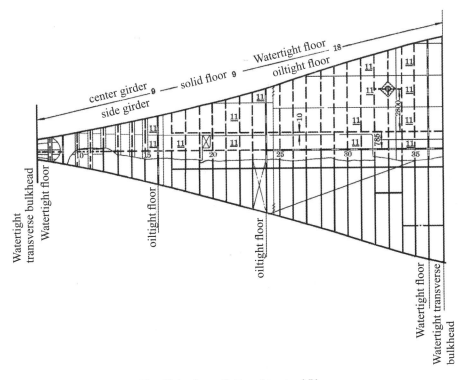

Fig. 10.3   Inner Bottom Structural Plan

The information it describe as following:

(1) The arrangements and thicknesses of inner bottom plates, the positions and sizes of stiffened panels and openings. The same lines in deck plans are applicable here.

(2) The positions, sizes and connection ways of members such as central girder, center keelson, side girder, bottom longitudinal, inner bottom longitudinal, watertight floor, solid floor, bracket floor, engine girder, etc.

New Words and Expressions

1. the basic structure plan 基本结构图
2. section ['sekʃn] n. 部分；断面；剖面图
3. sectional ['sekʃənl] adj. 剖面的；断面的；剖视的
4. the midship section plan 中横剖面图
5. frame [freɪm] n. 骨架；肋骨
6. projection [prə'dʒekʃn] n. 投影；投射；突起物
7. dashed lines 虚线
8. overlap [əuvəlæp] adj. 重叠的
9. double dots line 双点画线
10. side girder 舷侧纵桁
11. omit [ə'mɪt] vt. 省略；遗漏；删除；疏忽
12. platform ['plætfɔːm] n. 台；站台；平台；纲领
13. stiffened panels 加强面板
14. oblique lines 斜线
15. coaming 舱口围板
16. floor top 肋板上缘
17. applicable [ə'plɪkəb(ə)l] adj. 适当的；可应用的
18. center keelson 中内龙骨
19. bracket floor 框架肋板；组合肋板
20. engine girder 基座纵桁

Notes

1. The basic structure plan is composed of a longitudinal section view and several horizontal section views or sectional views, it is the basic drawings showing the structure of the hull.

   译：基本结构图由一个纵向剖面图和数个水平方向的剖面图或剖视图组成，是表示船体结构的基本图样。

2. The basic structure plan is a whole ship-like structural drawing, which is the basis for drawing other structural drawings and is also an instructional drawing during construction. Therefore, the basic structure plan of the hull must be correctly read and drawn.

 译：基本结构图是一张全船性的结构图样，它是绘制其他结构图样的依据，也是施工时的指导性图样。因此必须正确识读和绘制船体基本结构图。

3. The structural types ,sizes and connection ways of members in center plane. The projection profiles of these members are expressed by thin solid or dashed lines.

 译：中纵平面中构件的结构类型、尺寸和连接方式，这些构件的投影轮廓由细实线或细虚线表示。

Expanding reading

Midship Section Plan

  Midship section shows a cross section at the middle of the ship（Fig.10.4）. It illustrates the construction technique and materials. The plan gives the basic dimensions for most structural elements associated with the framing, planking and decking of the hull. In freighter it is always a cross-section of the hold closest to the midship. Some of the important data show include: principal dimensions; engine power and speed; data on classification; equipment numbers; maximum longitudinal bending moment.

  A local structural plan instead of a fully section may be introduced for different structures in a particular hold. Due to the symmetry of hull, it allows local members in different frame be drawn in another frame section plan by overlap projection. When this method is employed, thin double dot dash lines are used for visible outlines of these members and thin dashed lines for invisible outlines. A separate local plan is an alternative expression.

  Frame plans, if many, should be laid out from left to right, i. e. the frames near the stern will be drawn on the left and stem on right.

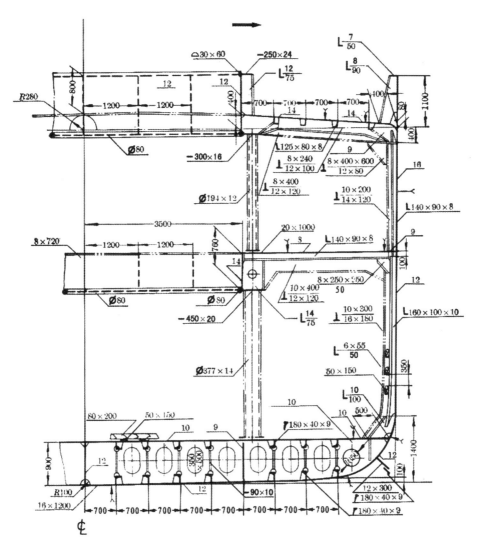

Fig. 10.4　Shell expansion plan

New Words and Expressions

1. midship section plan 中横剖面图
2. associate [əˈsəʊsieɪt] *v.* 相关, 联系
3. maximum longitudinal bending moment 最大纵向弯曲力矩
4. thin double dot dash lines 细双点画线
5. thin dashed lines 细虚线

Lesson 10 Basic Structural Plan / 85

Exercises

I. Answer the following questions according to the passage.

1. What is the function of the basic structure plan?
_____

2. What kind of information can you get from the basic structure plan?
_____

II. Practic these new words.

1. English to Chinese.

   section _____   sectional _____

   frame _____   ashed lines _____

   side girder _____   platform _____

2. Chinese to English.

   基本结构图 _____   中横剖面图 _____

   双点画线 _____   斜线 _____

   框架肋板 _____   基座纵桁 _____

III. Explain the nouns.

1. the basic structure plan.
_____

2. principal dimensions column.
_____

IV. Fill in the blanks with the proper words or expressions given below.

> center line, projection, dots dash lines, parallel to,
> omitted, dashed lines, overlap projected, arrangement

The longitudinal sectional view of the basic structural plan is a longitudinal sectional view obtained by cutting the hull through a plane _____ the hull center plane or a plane near the center plane. The cross-sectional view obtained by cutting the hull through the _____ plane of the hull, also known as the central longitudinal section, usually adopts this form. The longitudinal section view is generally arranged above the drawing surface, showing the _____ of the hull members in the length and depth of the ship and the connection of some longitudinal members. The central longitudinal section refers to a side view, i.e. profile along the center line. Lots of information can be found here:

(1) The structural types, sizes and connection ways of members in center plane. The _____ profiles of these members are expressed by thin solid or _____.

(2) Members between the center plane and the side of ship are _____, using thin double _____ for visible outlines and thin dashed lines for invisible outlines.

(3) Members located at the side of ship can be simplified as thick dot dash lines for side girders and web frames, thin dot dash lines for tween frames. The general frames can be _____.

(4) Members across the center line usually are cut here by using thick solid lines.

Ⅴ. Translation.

1. Translate the following sentences into Chinese.

(1) The structural types, sizes and connection ways of members in center plane. The projection profiles of these members are expressed by thin solid or dashed lines.

_____

(2) Members between the center plane and the side of ship are overlap projected, using thin double dots dash lines for visible outlines and thin dashed lines for invisible outlines.

_____

(3) Members located at the side of ship can be simplified as thick dot dash lines for side girders and web frames, thin dot dash lines for tween frames. The general frames can be omitted.

_____

(4) The arrangements and thicknesses of deck (platform) plates, the positions and sizes of stiffened panels and openings. Thin solid lines are used for seams and outlines of openings, while thin solid lines with oblique lines for stiffened panels.

_____

(5) The longitudinal and transverse bulkheads, coamings and stiffened beam knees. Usually, thick solid lines are used for visible bulkheads and coamings, rail lines for invisible watertight members, and thick dashed lines for invisible non-watertight members and knees.

2. Translate the short passage.
Inner Bottom Structural Plan

There are usually two forms of expression of the inner bottom structural plan: one is a cross-sectional view obtained by cutting the hull from a cut plane between the lowest deck and the bottom frame, so that the side structures that must be expressed can be clearly expressed, such as Fig.10.2 K-K section; the second is the method of stepwise sectioning by cutting the hull from a plan view near the upper edge of the bottom frame (the double bottom part uses a cutting plane near the upper surface of the inner bottom plate, and the single bottom part uses a cutting plane near the floor top) This view is relatively simple, such as J-J in Fig.10.2.

The information it describe as following:

(1) The arrangements and thicknesses of inner bottom plates, the positions and sizes of stiffened panels and openings. The same lines in deck plans are applicable here.

(2) The positions, sizes and connection ways of members such as central girder, center keelson, side girder, bottom longitudinal, inner bottom longitudinal, watertight floor, solid floor, bracket floor, engine girder, etc.

Module 5　Ship Production

Lesson 11　Ship Construction Process

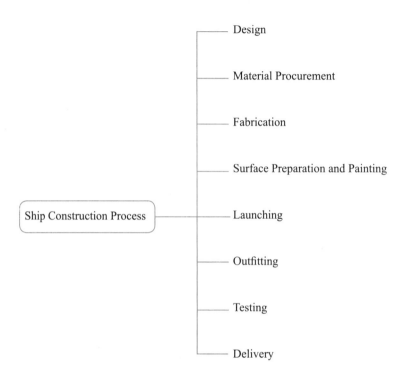

Background

A: Do you know the construction process of the ship?
B: The main steps in the ship construction process are design, material procurement, fabrication and so on. Next we will study in further detail.

Text reading

The construction of a ship is a highly technical and complicated process. Shipbuilding is performed for both military and commercial purposes. It is an international business, with major shipyards around the globe competing for a fairly limited amount of work.

Shipbuilding has changed radically since the 1980s. Formerly, most construction took place in a building or dock, with the ship constructed almost piece by piece from the ground up. However, advances in technology and more detailed planning have made it possible to construct the vessel in subunits or modules. Thus, the modules may be relatively easily connected. This process is faster, less expensive and provides better quality control. Further, this type of construction lends itself towards automation, not only saving money, but reducing the chemical and physical hazards.

The main steps in the ship construction process are as follows:

1. Design

The design considerations for various types of ships vary widely. In the design stage, not only should normal construction parameters be considered, but also the safety and health hazards associated with the construction process must be considered. In addition, environmental issues must be taken into account.The quality of production design is directly related to the progress and quality of ship production (Fig.11.1).

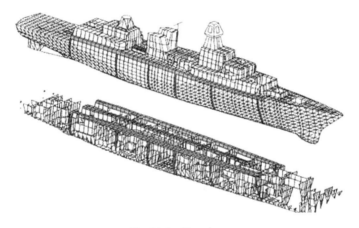

Fig. 11.1 Drawing

2. Material Procurement

The structural framework of most ships is constructed of various grades of mild and high-strength steel. Steel provides the formability, machinability and weldability required, combined with the strength needed for ocean-going vessels. These materials are required to perform a wide variety of functions,including the ship propulsion systems, back-up power, kitchens, pump stations for fuel transfer and so on.

3. Fabrication

The basic component of ship building is steel plate. The plates are cut, shaped, bent or otherwise manufactured into the desired shape specified by the design. Typically the plates are cut by an automatic flame cutting process to various shapes. These shapes may be then welded together to form I and T beams and other structural members.

The plates are then sent to fabrication shops, where they are joined into various units and sub-assemblies. In this moment, piping, electrical and other utility systems are assembled and integrated into the units. The units are assembled using automatic or manual welding or a combination of the two (Fig.11.2).

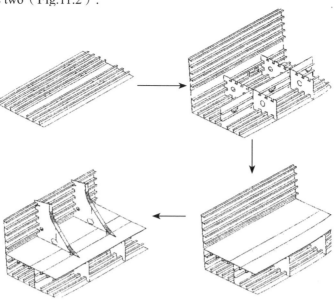

Fig. 11.2　Sectional Fabrication

　　The units or sub-assemblies are usually then transferred to an open-air area or lay down area where erection, or joining of assemblies, occurs to form even larger units or blocks. Here, additional welding and fitting occurs. Further, the units and welds must undergo quality-control inspections and testing. Those welds found defective must be removed by grinding or chiseling and then replaced.

　　The completed larger units are then moved to the dock, shipway or final assembly area. Here, the larger units are joined together to form the vessel. Again, much welding and fitting occur.

4. Surface Preparation and Painting

　　At the stage when even larger units or blocks are formed, the units are sand-blasted to ensure proper profiling, and painted. Paint may be applied by brush, roller or spray gun. Spraying is most commonly used. Painting is performed at almost every location in the shipyard. The type of paint needed for a certain application depends on the environment to which the coating will be exposed (Fig.11.3).

Fig. 11.3　Surface Preparation and Painting

5. Launching

Once the hull is structurally complete and watertight, the vessel is launched. This may involve sliding it into the water from the shipway on which it was constructed, flooding of the dock in which it was constructed or lowering the vessel into the water. Launching is almost always accompanied by great celebration.

6. Outfitting

After the ship is launched, it enters the outfitting stage. A large amount of time and equipment are required. The work includes the fitting of cabling and piping, the furnishing of galleys and accommodations, insulation work, installation of electronic equipment and navigation aids and installation of propulsion and ancillary machinery. This work is performed by a wide variety of skilled trades (Fig.11.4).

Fig. 11.4　Outfitting

7. Testing

After completion of the outfitting phase, the ship undergoes both dock and sea trials, during which all the ship's systems are proved to be fully functional and operational. Once testing is performed, the ship is sent to sea for a series of fully operational tests and sea trials before the ship is delivered to its owner.

8. Delivery

Finally, after all testing work is performed, the ship is delivered to the customer.

New Words and Expressions

1. military ['mɪlətri] *adj.* 军事的
2. commercial [kəˈmɜː(ə)ʃl] *adj.* 贸易的，商业的
3. formerly [ˈfɔːrmərli] *adv.* 以前，从前
4. automation [ˌɔːtəˈmeɪʃ(ə)n] *n.* 自动化
5. hazard [ˈhæzərd] *n.* 危险，危害
6. parameter [pəˈræmɪtər] *n.* 参数
7. material procurement 材料采购

8. formability [ˌfɔːməˈbɪlɪti] n. 可成形性

9. propulsion [prəˈpʌlʃn] n. 推动力，推进

10. automatic flame cutting process 火焰自动切割工艺

11. sectional fabrication 分段装配

12. grinding [ˈɡraɪndɪŋ] v. 磨碎，磨光

13. chiseling [ˈtʃɪzlɪŋ] v. 凿开（chise 的 ing 形式）

14. ancillary machinery 辅助机械

15. sand-blasted 喷砂的

16. spraying [ˈspreɪɪŋ] n. 喷涂

17. furnishing [ˈfɜːnɪʃɪŋ] n. 家具

18. navigation aids [ˌnævɪˈɡeɪʃn eɪdz] n. 导航设备

19. delivery [dɪˈlɪvəri] n. 交付

Notes

1. In the design stage, not only should normal construction parameters be considered, but also the safety and health hazards associated with the construction process must be considered.

   译：在设计阶段，不仅应考虑正常的施工参数，而且还必须考虑与施工过程相关的安全和健康危害。

2. Steel provides the formability, machinability and weldability required, combined with the strength needed for ocean-going vessels.

   译：钢提供了所需的成形性、可加工性和焊接性，以及远洋船舶所需的强度。

3. The work includes the fitting of cabling and piping, the furnishing of galleys and accommodations, insulation work, installation of electronic equipment and navigation aids and installation of propulsion and ancillary machinery.

   译：具体工作包括安装电缆和管道、提供厨房和住宿、绝缘工程、安装电子设备和助航设备以及安装推进和辅助机械。

Expanding reading

Brief Introduction of Testing

The operation and test stage of construction assesses the functionality of installed components and systems. At this stage, systems are operated, inspected and tested. If the systems fail the tests for any reason, the system must be repaired and retested until it is fully operational. All piping systems on board the ship are pressurized to locate leaks that may exist in the system. Tanks also need structural testing, which is accomplished by filling the tanks with fluids (i.e., salt water or fresh water) and inspecting for structural stability. Ventilation, electrical and many other systems are tested. Most system testing and operations occur while the ship is docked at hoverport. However, there is an increasing trend to perform testing at earlier stages of construction (e.g., preliminary testing in the production shops). Performing tests at earlier stages of construction makes it easier to fix failures because of the increased accessibility to the systems, although complete systems tests will always need be done on board (Fig.11.5).

Fig. 11.5 Testing

New Words and Expressions

1. functionality [ˌfʌŋkʃəˈnæləti] *n.* 设计目的，功能
2. install [ɪnˈstɔːl] *v.* 安装，设置
3. pressurized [ˈpreʃəraɪzd] *adj.* 加压的，增压的
4. leak [liːk] *n.* 漏洞，裂缝
5. accomplish [əˈkʌmplɪʃ] *v.* 完成
6. ventilation [ˌventɪˈleɪʃn] *n.* 通风设备
7. hoverport [ˈhɒvəˌpɔːt] *n.* 码头
8. preliminary testing 初步测试
9. production shop 生产车间

Exercises

I. Answer the following questions according to the passage.

1. What are the steps of ship building process?

2. What equipment does the outfitting work include?

II. Practic these new words.

1. English to Chinese.

military _____   parameter _____

commercial _____   automation _____

formability _____   propulsion _____

2. Chinese to English.

分段装配 _____   辅助机械 _____

喷砂的 _____   导航设备 _____

低碳钢 _____   材料采购 _____

III. Explain the nouns.

1. delivery.
   _____

2. Surface Preparation.
   _____

IV. Fill in the blanks with the proper words or expressions given below.

> quality control, piece by piece, Formerly,
> amount of, Shipbuilding, modules, international

　　The construction of a ship is a highly technical and complicated process. _____ is performed for both military and commercial purposes. It is an _____ business, with major shipyards around the globe competing for a fairly limited _____ work.

　　Shipbuilding has changed radically since the 1980s. _____ , most construction took place in a building or dock, with the ship constructed almost _____ from the ground up. However, advances in technology and more detailed planning have made it possible to construct the vessel in subunits or modules. Thus, the _____ may be relatively easily connected. This process is faster, less expensive and provides better _____. Further, this type of construction lends itself towards automation, not only saving money, but reducing the chemical and physical hazards.

V. Translation.

1. Translate the following sentences into Chinese.

(1) Formerly, most construction took place in a building or dock, with the ship constructed almost piece by piece from the ground up.

(2) Further, this type of construction lends itself towards automation, not only saving money, but reducing the chemical and physical hazards.

(3) The quality of production design is directly related to the progress and quality of ship production.

(4) After completion of the outfitting phase, the ship undergoes both dock and sea trials, during which all the ship's systems are proved to be fully functional and operational.

(5) This may involve sliding it into the water from the shipway on which it was constructed, flooding of the dock in which it was constructed or lowering the vessel into the water.

2. Translate the short passage.

The basic component of ship building is steel plate. The plates are cut, shaped, bent or otherwise manufactured into the desired shape specified by the design. Typically the plates are cut by an automatic flame cutting process to various shapes. These shapes may be then welded together to form I and T beams and other structural members.

The plates are then sent to fabrication shops, where they are joined into various units and sub-assemblies. In this moment, piping, electrical and other utility systems are assembled and integrated into the units. The units are assembled using automatic or manual welding or a combination of the two.

Lesson 12　Fabrication and Welding of Hull (1)

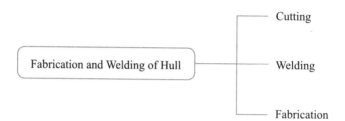

Background

A: Do you know how many methods for the fabrication and welding of hull?

B: In general, two methods for their fabrication and welding are available: pyramid method and converse method, Next we will study in further detail.

Text reading

Before the fabrication of hull, it is necessary to cut the components according to the relevant design drawings, then the cutting steel plates are welded together, and finally the fabrication is performed.

1. Cutting

The "fabrication line" of the shipyard stars in the steel storage area. Here, large steel plates of various strengths, sizes and thicknesses are stored and readied for construction.

The steel is then blasted with abrasive and primed with a construction primer that preserves the steel during the various phases of construction. The steel plate then is transported to a construction facility. Here the steel plate is cut by automatic burners to the desired size. The resulting strips are then welded together to form the structural components of the vessel (Fig.12.1).

Fig. 12.1   Cutting

2. Welding

Welding for hull is one of the primary workmanship in shipbuilding, and so the quality, working efficiency and cost of shipbuilding directly rely on it. Broadly speaking, steel is an excellent material for shipbuilding purposes, and the choice of welding electrode is the utmost importance in all welding applications during construction. The standard goal is to obtain a weld with equivalent strength characteristics to that of the base metal. Since minor flaws are likely to occur in production welding, welds are often designed and welding electrodes chosen to produce welds with properties in excess of those of the base metal.

Shipyard welding processes, or more specifically fusion welding, is performed at nearly every location in the shipyard environment. The process involves joining metals by bringing adjoining surfaces to extremely high temperatures to be fused together with a filler material. A heat source is used to heat the edges of the joint, permitting them to fuse with molten weld fill metal.

The required heat is usually generated by an electric arc or a gas flame. Shipyards choose the type of welding process based on customer specifications, production rates and a variety of operating constraints including government regulations. Standards for military vessels are usually more stringent than commercial vessels (Fig.12.2).

Fig.12.2   Welding

3. Fabrication

This workmanship mainly includes:

A. Part fabrication—to assemble different kinds of members into a whole part after their processing.

B. Block or section fabrication—to assemble different parts into a block or section.

C. Berth or slipway fabrication—to assemble a few blocks and sections into an integral hull.

(1) Division of Sections

Division of sections is of great importance, for not only the strength of hull itself but also the convenience and reasonableness of operation as well as the production procedure, lifting capacity and arrangement of the working site have got to be taken into our account. To make the matter worse, the aforementioned factors are frequently contrary to one another in one way or another. Therefore, it is not rare for us to rack our brains to find ways and means available (Fig.12.3).

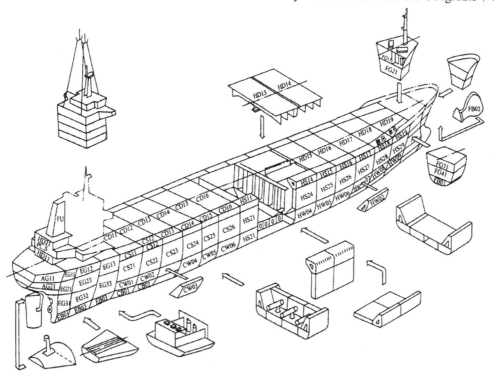

Fig. 12.3 Division of Sections

(2) Fabrication and Welding of Parts

The fabrication and welding of parts, such as frames, beams, girders and seatings, to be performed on a platform in term of the concerned drawings.

(3) Fabrications and Welding of Blocks and Sections

The fabrication and welding of blocks, such as side, bottom, deck and superstructure, be conducted on a jig. In respect to sections, two methods for their fabrication and welding are available:

① Pyramid Method

The practice of this method is as follows;First, lift a bottom block onto a berth or a slipway and take it both as a base and a jig.Then, assemble side blocks and bulkhead blocks Still then,assemble the deck block. By the pyramid method,all are done bit by bit through welding （Fig.12.4, Fig.12.5）.

Fig.12.4　Pyramid Method　　Fig.12.5　Converse Method

② Converse Method

This method is good for bow section , stern section, deckhouse and superstructure , the main characteristic of which is to regard the deck block on a berth or a slipway both as a base and a jig.Thus framing blocks , bulkhead blocks and side girders will be erected on the base of a deck block.

New Words and Expressions

1. fabrication [ˌfæbrɪˈkeɪʃn] *n.* 制造，建造，装配
2. assemble [əˈsembəl] *v.* 装配
3. construction primer 施工底漆
4. phase [feɪz] *n.* 阶段
5. automatic burner 自动燃烧器
6. electrode [ɪˈlektrəʊd] *n.* 电极
7. equivalent [ɪˈkwɪvələnt] *adj.* 相等的，相同的
8. minor flaw 小瑕疵
9. fusion [ˈfju:ʒ(ə)n] *n.* 融合
10. fuse together *v.* 融为一体
11. government regulation 政府规章
12. part fabrication 部装
13. block or section fabrication 分段或总段装配
14. berth or slipway fabrication 船台装配
15. platform [ˈplætfɔ:m] *n.* 平台
16. jig [dʒɪg] *n.* 胎架
17. working site 现场
18. pyramid method 正装法
19. converse method 反装法

20. framing block 肋骨框架分段

Notes

1. The steel is then blasted with abrasive and primed with a construction primer that preserves the steel during the various phases of construction.

    译：然后用磨料对钢进行喷砂处理，并涂上施工底漆，以在施工的各个阶段保护钢材。

2. Broadly speaking, steel is an excellent material for shipbuilding purposes, and the choice of welding electrode is the utmost importance in all welding applications during construction.

    译：从广义上讲钢是造船用的优良材料，在施工过程中，焊接电极的选择是所有焊接应用的关键。

3. The process involves joining metals by bringing adjoining surfaces to extremely high temperatures to be fused together with a filler material.

    译：这一过程包括通过使相邻表面达到极高的温度与填充材料熔融在一起来连接金属。

Expanding reading

The digital shipyard

Over recent years, shipbuilders have been steadily moving away from their old production methods to embrace "smart" manufacturing approaches and bring streamlined, data-rich efficiency to the design and build process.

The next generation, digitised and intelligent shipyard not only promises cheaper and more efficient design and construction, but should also drive down the cost of ownership too. The key is creating a digital thread, a synchronised body of information that encompasses the entire supply chain, and builds into what has been called a "single version of the truth" that governs everything from conception, design and construction, to upgrades and modifications throughout the vessel's in-service life.

Shipbuilding 4.0: An automation and data-exchange revolution

Under the traditional shipbuilding model, separate and discrete unit workflows were the order of the day, with the overall design broken down into a series of individual task areas, each unconnected to the next, and essentially without any level of direct interaction. As work progressed, plans would travel backwards and forwards between the individual teams so that changes and developments could be reconciled manually, which obviously extends the build time, while also multiplying the potential for mistakes or omissions to creep in. In addition, denied the immediacy of real-time collaboration, any cross-fertilization of ideas between the different teams was also virtually impossible.

Move forward to what some have dubbed the "shipbuilding 4.0" model in reference to the adoption of the so-called "industry 4.0" automation and data-exchange revolution that is sweeping manufacturing in general, and those issues become a thing of the past. The digital shipyard replaces the old isolated, disparate technology platforms and their compartmentalized data, with united state-of-the-art planning tools, and a single common repository of design data that is always current, and available to anyone who needs it（Fig.12.6）.

Fig.12.6　Digital Shipbuilding

New Words and Expressions

1. the digital shipyard 数字化船厂
2. embrace [ɪmˈbreɪs] v. 欣然接受
3. streamlined [ˈstriːmlaɪnd] v. 精简
4. data-rich efficiency 更高的效率
5. synchronise [ˈsɪŋkrənaɪz] vi. 同时发生
6. automation and data-exchange revolution 自动化和数据交换革命
7. potential [pəˈtenʃl] n. 可能性，潜在性
8. cross-fertilization [ˌkrɒsˌfɜːtəlaɪˈzeɪʃn] 相互融合
9. compartmentalize [kəmˌpɑːrtˈmentəlaɪz] v. 分隔，隔开，划分
10. repository [rɪˈpɑːzətɔːri] n. 数据库

Exercises

I. Answer the following questions according to the passage.

1. How to protect the steel plate during transportation and processing?

2. How to weld two pieces of metal together?

Ⅱ. Practic these new words.

1. English to Chinese.

   construction primer _____   fabrication _____

   automatic burner _____   assemble _____

   fuse together _____

   construction _____

2. Chinese to English.

   部装 _____   反装法 _____

   船台装配 _____   正装法 _____

   现场 _____   总段装配 _____

Ⅲ. Explain the nouns.

1. pyramid method.
   _____

2. converse method.
   _____

Ⅳ. Fill in the blanks with the proper words or expressions given below.

> a variety of, permitting, welding, shipyard,
> fused together, Standards, generated by

Shipyard welding processes, or more specifically fusion _____, is performed at nearly every location in the _____ environment. The process involves joining metals by bringing adjoining surfaces to extremely high temperatures to be _____ with a filler material. A heat source is used to heat the edges of the joint, _____ them to fuse with molten weld fill metal.

The required heat is usually _____ an electric arc or a gas flame. Shipyards choose the type of welding process based on customer specifications, production rates and _____ operating constraints including government regulations. _____ for military vessels are usually more stringent than commercial vessels.

Ⅴ. Translation.

1. Translate the following sentences into Chinese.

(1) Here the steel plate is cut by automatic burners to the desired size. The resulting strips are then welded together to form the structural components of the vessel.

_____

(2) Shipyard welding processes, or more specifically fusion welding, is performed at nearly every location in the shipyard environment.

_____

(3) Division of sections is of great importance, for not only the strength of hull itself but also the convenience and reasonableness of operation as well as the production procedure, lifting capacity and arrangement of the working site have got to be taken into our account.

_____

(4) To make the matter worse, the aforementioned factors are frequently contrary to one another in one way or another.Therefore,it is not rare for us to rack our brains to find ways and means available.

_____

(5) The fabrication and welding of blocks, such as side, bottom, deck and superstructure, be conducted on a jig.

_____

2. Translate the short passage.

(1) Pyramid Method

　　The practice of this method is as follows;First, lift a bottom block onto a berth or a slipway and take it both as a base and a jig.Then, assemble side blocks and bulkhead blocks Still then,assemble the deck block. By the pyramid method,all are done bit by bit through welding.

(2) Converse Method

　　This method is good for bow section, stern section, deckhouse and superstructure, the main characteristic of which is to regard the deck block on a berth or a slipway both as a base and a jig. Thus framing blocks, bulkhead blocks and side girders will be erected on the base of a deck block.

_____

_____

_____

Lesson 13　Fabrication and Welding of Hull (2)

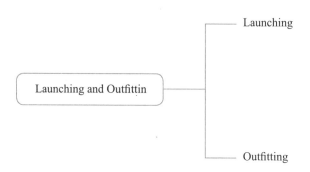

Background

A: What do you know about the outfitting?

B: Usually, after launching, the ship is berthed in a fitting-out basin for completion.The main machinery, together with auxiliaries, piping systems, deck gear, lifeboats, accommodation equipment, pumping systems, and rigging are installed on board. Next we will study in further detail.

Text reading

1. Launching

　　The launching methods of the ships can be divided into three types: gravity launching, floating launching, mechanized launching. Gravity launching is suitable for most ships. Floating launching is suitable for very large ships. Mechanized launching is mainly suitable for small and medium-sized ships（Fig.13.1）.

Fig. 13.1　Floating launching　　Fig. 13.2　Mechanized launching

Gravity launching is divided into longitudinal oiling slide launching, longitudinal ball launching and side oiling slide launching.This is also the main gravity launching method.Next,we mainly introduce gravity launching.

Apart from certain small craft built on inland waterways, which are launched sideways, the great majority of ships are launched stern first from the building berth. Standing structures called ways, constructed of concrete and wooden blocks, spaced about one-third of the vessel's beam apart, support the ship under construction. The slope of the standing ways-which are often cambered (slightly curved upward toward the middle or slightly curved downward toward the ends) in the fore and aft direction-ranges; ways extend from a position near the water.

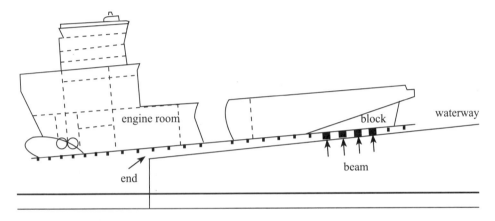

Fig. 13.3　Gravity launching

During construction the ship is supported by at least one line of blocks under the keel, with side supports and shores as necessary. As the vessel nears completion, the standing ways are built under it, the sliding ways are superimposed, and the cradle is built up.

The weight of the vessel is transferred to the standing ways. The full weight must not be supported by the ways for too long because the thickness of lubricant would be reduced by squeezing and its properties would be adversely affected. It is common to fit launching triggers which when released at the moment of launching, permit the ship to move over the standing ways.

As a vessel moves down the ways, the forces operating are: its weight acting down through the centre of gravity, the upward support from the standing ways, and the buoyancy of the water. As it travels further, the buoyancy increases and the upthrust of the ways decreases, with the weight remaining constant. As the centre of gravity passes the after end of the standing ways, the moment of the weight about the end of the ways trends to tip the ship stern first. At this position and for some time later, it is essential that the moment of buoyancy be greater than the moment of weight about the after end of the ways, thus giving a moment to keep the forward end of the sliding ways on the standing ways; otherwise there would be concentration of weight at the end of the ways, causing excessive local pressure. Calculations are made to determine the most important factors in lifts, the difference between weight and buoyancy when the stern lifts, the existence of the ways to ensure that the cradle will not drop off the end of the standing ways.

2. Outfitting

After launching, the ship is berthed in a fitting-out basin for completion. The main machinery, together with auxiliaries, piping systems, deck gear, lifeboats, accommodation equipment, pumbing systems, and rigging are installed on board, along with whatever insulation and deck coverings are necessary. Fitting out may be a relatively minor undertaking, as with a tanker or a bulk carrier, but in the case of a passenger vessel, the work will be extensive. Although fitting-out operations are diverse and complex, as with hull construction here are four main divisions: ① collection and grounding of the specified components, ② installation of components according to schedule, ③ connection of components to appropriate piping and/or wiring systems, and ④ testing of completed systems (Fig.13.4).

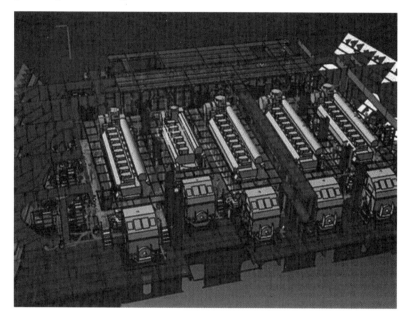

Fig. 13.4　Outfitting

In the early 1970s, the tendency in planning was to divide the ship into sections, listing the quantities of components required and times of delivery. Drawings necessary for each section are prepared and these specify the quantities of components required. A master schedule is complied, specifying the sequences and target dates for completion and testing of each component system. This schedule is used to marshal and synchronize fitting work in the different sections and compartments.

New Words and Expressions

1. gravity launching 重力式下水
2. floating launching 漂浮式下水
3. mechanized launching 机械式下水

4. longitudinal oiling slide launching 纵向涂油滑道下水
5. longitudinal ball launching 纵向钢珠滑道下水
6. superimpose [ˌsuːpərɪmˈpəʊz] vt. 添加；重叠；附加；安装
7. launching cradle 下水架，发射架
8. lubricant [ˈluːbrɪkənt] n. 润滑剂
9. standing way 滑道
10. side support 边墩
11. trigger [ˈtrɪɡə(r)] n. 下水扳机
12. upthrust [ˈʌpθrʌst] n. 向上反力
13. a fitting-out basin 舾装泊位
14. plumbing system 污水系统
15. rigging [ˈrɪɡɪŋ] n. 索具
16. insulation [ˌɪnsjuˈleɪʃn] n. 绝缘
17. deck covering 甲板敷料
18. target dates 预定日期
19. marshal [ˈmɑːʃ(ə)l] vt. 调度
20. synchronize [ˈsɪŋkrənaɪz] vt. 协调

Notes

1. The launching methods of the ships can be divided into three types: gravity launching, floating launching, mechanized launching.

   译：船舶下水方式可分为重力式下水、漂浮式下水和机械化下水三种。

2. Apart from certain small craft built on inland waterways, which are launched sideways, the great majority of ships are launched stern first from the building berth.

   译：除了某些建造在内河航道上的小艇是侧向下水外，绝大多数船舶是船尾先从船坞下水。

3. As a vessel moves down the ways, the forces operating are: its weight acting down through the centre of gravity, the upward support from the standing ways, and the buoyancy of the water.

   译：当一艘船向下移动时，作用的力是：通过重心向下的重力、滑道向上支撑力和水的浮力。

4. The main machinery, together with auxiliaries, piping systems, deck gear, lifeboats, accommodation equipment, pumping systems, and rigging are installed on board, along with whatever insulation and deck coverings are necessary.

   译：主机、辅机、管道系统、甲板设备、救生艇、住宿设备、抽水系统和索具，连同必要的绝缘材料和甲板敷料一起安装在船上。

Expanding reading

Sea Trials

Taking a brand new ship to sea for the first time is naturally quite an event after so many months of preparation and hard work, and the prevailing spirit is one of cheerful optimism, especially if the hopes of good weather have materialized. As soon as the trials party is assembled on board, the pilot, who has probably been waiting in the wheelhouse for some time, orders the gangway away and the mooring ropes cast off and takes the ship out of harbour. Sea trial programmes vary somewhat to suit owners particular requirements, but most have several trials in common which will be briefly described.

Before the ship has proceeded very far out to sea and while she is still in shallow water, the engines will probably be stopped in order to carry out anchor-heaving trials on the wind lass; after which the magnetic compass will be adjusted by a specialist known as a compass adjuster, who will carry out his work by a process called 'swinging ship'. Let us assume that the ship now shapes course for the measured mile, to run a series of speed trials. On the way steering trials will be carried out in order to check the time it takes for the rudder to be put from hard-over one side to hard-over the other. Manoeuvring trials may also be carried out to determine the ship's turning circle and the time and distance from engines full ahead to full astern (Fig.13.5).

Fig.13.5 Sea Trials

New Words and Expressions

1. sea trials 试航
2. prevailing [prɪˈveɪlɪŋ] *adj.* 普遍的，盛行的
3. wheelhouse [ˈwiːlhaʊs] *n.*(船上的) 操舵室，驾驶室
4. gangway [ˈɡæŋweɪ] *n.* 舷门，舷梯，步桥，跳板
5. mooring ropes cast off 系泊缆绳脱落

Lesson 13　Fabrication and Welding of Hull (2)

6. anchor-heaving trials 锚泊试验
7. magnetic compass 磁罗经
8. manoeuvring trials 操纵试验

Exercises

Ⅰ. **Answer the following questions according to the passage.**

1. What does the launching include?

   _____

2. What does the gravity launching include?

   _____

Ⅱ. **Practic these new words.**

1. English to Chinese.

   gravity launching _____　　floating launching _____

   launching cradle _____　　trigger _____

   rigging _____　　insulation _____

2. Chinese to English.

   机械式下水 _____　　边墩 _____

   滑道 _____　　舾装泊位 _____

   污水系统 _____　　甲板敷料 _____

Ⅲ. **Explain the nouns.**

1. the outfitting.

   _____

2. the launching.

   _____

IV. Fill in the blanks with the proper words or expressions given below.

> buoyancy, determine, the upward support, the upthrust,
> trends to, the sliding ways, drop off, centre of gravity

As a vessel moves down the ways, the forces operating are: its weight acting down through the centre of gravity, _____ from the standing ways, and the buoyancy of the water. As it travels further, the buoyancy increases and _____ of the ways decreases, with the weight remaining constant. As the _____ passes the after end of the standing ways, the moment of the weight about the end of the ways _____ tip the ship stern first. At this position and for some time later, it is essential that the moment of _____ be greater than the moment of weight about the after end of the ways, thus giving a moment to keep the forward end of _____ on the standing ways; otherwise there would be concentration of weight at the end of the ways, causing excessive local pressure. Calculations are made to _____ the most important factors in lifts, the difference between weight and buoyancy when the stern lifts, the existence of the ways to ensure that the cradle will not _____ the end of the standing ways.

V. Translation.

1. Translate the following sentences into Chinese.

(1) Gravity launching is suitable for most ships. Floating launching is suitable for very large ships. Mechanized launching is mainly suitable for small and medium-sized ships.

_____

(2) Gravity launching is divided into longitudinal oiling slide launching, longitudinal ball launching and side oiling slide launching.

_____

(3) Standing structures called ways, constructed of concrete and wooden blocks, spaced about one-third of the vessel's beam apart, support the ship under construction.

_____

(4) During construction the ship is supported by at least one line of blocks under the keel, with side supports and shores as necessary.

_____

(5) The full weight must not be supported by the ways for too long because the thickness of lubricant would be reduced by squeezing and its properties would be adversely affected.

_____

2. Translate the short passage.

Outfitting

After launching, the ship is berthed in a fitting-out basin for completion. The main machinery, together with auxiliaries, piping systems, deck gear, lifeboats, accommodation equipment, pumbing systems, and rigging are installed on board, along with whatever insulation and deck coverings are necessary. Fitting out may be a relatively minor undertaking, as with a tanker or a bulk carrier, but in the case of a passenger vessel, the work will be extensive. Although fitting-out operations are diverse and complex, as with hull construction here are four main divisions: ① collection and grounding of the specified components, ② installation of components according to schedule, ③ connection of components to appropriate piping and/or wiring systems, and ④ testing of completed systems.

In the early 1970s, the tendency in planning was to divide the ship into sections, listing the quantities of components required and times of delivery. Drawings necessary for each section are prepared and these specify the quantities of components required. A master schedule is complied, specifying the sequences and target dates for completion and testing of each component system. This schedule is used to marshal and synchronize fitting work in the different sections and compartments.

Module 6　Ship Survey and Quality Management

Lesson 14　Survey

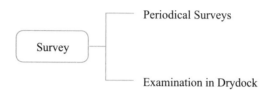

Background

A: Do you know what the survey include?

B: It includes periodical surveys, examination in drydock, hull surveys of very large crude carriers and so on. Next we will study in further detail.

Text reading

In common with all machinery a ship requires regular overhaul and maintenance. The particularly severe operating conditions for an almost all-steel structure necessitate constant attention to the steelwork. The operations of berthing, cargo loading and discharge, constant immersion in sea water and the variety of the extremes encountered all take their tolls on the structure and its protective coatings. The classification societies have requirements for examination or survey of the ship at set periods throughout its life. The nature and extent of the survey increases as the ship becomes older.

1. Periodical Surveys

All ships must have an annual survey, which is carried out by a surveyor employed by the classification society. This survey should preferably take place in a drydock (Fig.14.1), but the period between in dock surveys may be extended up to 2.5 years, such an extension is permitted where the ship is coated with a high resistance paint and an approved automatic impressed current cathodic protection system is fitted, in-water surveys are permitted for ships which are less than 10 years old and greater than 38 m in breadth and have the paint and cathodic protection systems already referred to. Special surveys of more rigorous nature are required every 4 years. Continuous surveys are permitted where all the various hull compartments are examined in rotation over a period of 5 years between consecutive examinations.

Fig. 14.1 drydock

During an annual survey the various closing appliances on all hatchways and other hull opening through which water might enter must be checked to be in efficient. Water-clearing arrangements, such as scuppers and bulwark freeing ports, must also operate satisfactorily. Guard rails, lifelines and gangways are also examined.

When surveyed in drydock the hull plating is carefully examined for any signs of damage. The stern frame and rudder are also examined for cracks, etc. The wear in the rudder and propeller shaft bearings is also measured.

The fire protection, detection and extinguishing arrangements for passenger ships are examined every year and for cargo ships every two years (Fig.14.2).

Fig. 14.2 fire protection

For a special survey, the requirements of the annual survey must be met together with examinations. A detailed examination of structure by removing covers and linings may be made. Metal thicknesses at any areas showing wastage may have to be checked. The double-bottom and peak tanks must be tested by filling to the maximum service head with water. The decks, casings and superstructures, together with any areas of discontinuity, must be examined for cracks or signs of failure. All escape route from occupied or working spaces must be checked. Emergency communications to the machinery space and the auxiliary steering position from the bridge must also be proved.

For tankers, additional special survey requirements include the inspection of all cargo tanks and cofferdam spaces. Cargo tank bulkheads must be tested by filling to the top of the hatchway of all or alternate tanks. The greater the age of a ship the greater will be the detail of examination and testing of suspect or corrosion-prone spaces.

Liquefied gas tankers have requirements for annual surveys, as mentioned earlier and several additional items. All tanks, cofferdams, pipes, etc. , must be gas freed before survey. Where the maximum vapour pressure in the tanks is 0.7 bar or less the inner tank surfaces are to be examined. In addition, the tanks must be water tested by a head of 2.45 m above the top of the tank. All tank level devices, gas detectors, inertia arrangements, etc. , must be proved to be operating satisfactorily. The special survey requirements are as previously stated, together with the examination internally and externally where possible of all tank areas. Tank mountings, supports, pipe connections and deck sealing arrangements must also be checked. Samples of insulation, where fitted, must be removed and the plating beneath examined. Pressure-relief and vacuum valves must be proved to be efficient. Refrigeration machinery must be examined.

2. Examination in Drydock

The drydock of a ship provides a rare opportunity for examination of the underwater areas of a ship. Every opportunity should therefore be taken by the ship's staff, the shipowner and the classification society to examine the ship thoroughly. Some of the more important areas are now listed.

(1) Shell Plating

The shell plating must be thoroughly examined for any corrosion of welds, damage, distortion and cracks at openings or discontinuities. Any hull attachments such as lugs, bilge keels, etc., must be checked for corrosion, security of attachment and any damage. All openings for grids and sea boxes must also be examined (Fig.14.3).

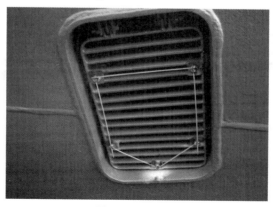

Fig. 14.3　Grids

（2）Cathodic Protection Equipment

Sacrificial anodes should be checked for security of attachment to the hull and the degree of wastage that has taken place. With impressed current systems the anodes and reference anodes must be checked, again for security of attachment. The inert shields and paintwork near anodes should be examined for any damage or deterioration.

（3）Rudder

The plating and visible structure of the rudder should be examined for cracks and any distortion. The drain plugs should be removed to check for the entry of any water. Pintle or bearing weardown and clearances should be measured and the security of the rudder stock coupling bolts and any nuts should be ensured.

（4）Stern frame

The surface should be carefully checked for cracks, particularly in the areas where a change of section occurs or large bending moments are experienced.

（5）Propeller

The cone should be checked for security of attachment and also the rope guard. The blades should be examined for corrosion and cavitation damage, and any cracks or damage to the blade tips. It is usual to examined any tail shaft seals and also measure the tail shaft wear down.

New Words and Expressions

1. overhaul [ˌəʊvəˈhɔːl] *n.* 分解检查，大修
2. maintenance [ˈmeɪntənəns] *n.* 维护，维修
3. berth [bɜːθ] *v.* 停泊
4. surveyor [səˈveɪə(r)] *n.* 测量员；检查员；调查员
5. gangway [ˈɡæŋweɪ] *n.* 舷梯；跳板；进出通路
6. stern frame *n.* 艉柱
7. casing [ˈkeɪsɪŋ] *n.* 框；壳；罩；套
8. cofferdam space 隔离舱
9. hatchway [ˈhætʃweɪ] *n.* 舱口
10. corrosion-prone spaces 易腐蚀空间
11. refrigeration machinery 制冷机械
12. shell plating 外壳镀层
13. distortion [dɪˈstɔːʃ(ə)n] *n.* 扭曲，变形
14. attachment [əˈtætʃmənt] *n.* 附件；连接物
15. grids [ɡrɪdz] *n.* 格栅
16. cathodic protection equipment 阴极保护设备
17. wastage [ˈweɪstɪdʒ] *n.* 损耗；损耗量
18. propeller [prəˈpelə(r)] *n.* 螺旋桨；推进器
19. cavitation [ˌkævəˈteɪʃən] *n.* 气蚀

Notes

1. The drydock of a ship provides a rare opportunity for examination of the underwater areas of a ship. Every opportunity should therefore be taken by the ship's staff, the shipowner and the classification society to examine the ship thoroughly.

   译：船舶干船坞为检验船舶水下区域提供了难得的机会。因此，船舶的工作人员、船东和船级社应利用一切机会彻底检查船舶。

2. The shell plating must be thoroughly examined for any corrosion of welds, damage, distortion and cracks at openings or discontinuities. Any hull attachments such as lugs, bilge keels, etc., must be checked for corrosion, security of attachment and any damage.

   译：必须彻底检查壳板是否有任何焊接腐蚀、损坏、变形和开口或不连续处的裂纹。任何船体附件，如吊耳、舭龙骨等，必须检查是否有腐蚀、附件的安全性和任何损坏。

3. The blades should be examined for corrosion and cavitation damage, and any cracks or damage to the blade tips. It is usual to examined any tail shaft seals and also measure the tail shaft wear down.

   译：应该检查叶片是否有腐蚀和气蚀损伤，以及叶片尖端是否有任何裂纹或损伤。通常检查每个艉轴密封件，并测量艉轴磨损。

Expanding reading

Hull Surveys of Very Large Crude Carriers

The very size of these ships necessitates considerable planning and preparation prior to any survey. Large amounts of staging is necessary to test its structure. Good lighting, safe access and some means of communication are also required. Surveys are often undertaken at with the gas freeing of the tanks being one of the main problems. In-water surveys of the outer hull are also done. Some thought at the design stage of the ship should enable the stern bush, pintle and rudder bush clearances to be measured in the water. Provision should also exist for unshipping the propeller in the water. Anodes should be bolted to the shell and therefore easily replaced. Blanks for sealing off inlets should be carried by the ship, to enable the overhaul of ship side valves. The frame markings should be painted on the outside of the ship at the weather deck edge to assist in identifying frames and bulkheads. An in-water survey plan should be prepared by the shipbuilder. The hull plating surface must be clean prior to survey. This can be achieved by the use of rotary hand-held brushes which may be hydraulically or pneumatically powered. In-water cleaning of the hull is possible, with drivers using three brushes or specially designed boats with long rotating brushes attached. (Fig.14.4)

Fig.14.4　Brush

Lesson 14 Survey

New Words and Expressions

1. Very Large Crude Carriers 超级油轮
2. stern bush 艉轴套
3. pintle [ˈpɪntəl] 舵销
4. rudder bush 舵钮衬套
5. unship [ʌnˈʃɪp] *v.* 卸货，下船
6. anode [ˈænəud] *n.* 阳极
7. prior [ˈpraɪər] *adj.* 先前的，优先的
8. hydraulically [haɪˈdrɔlɪkli] *adv.* 应用液压
9. pneumatically [njuːˈmætɪkəli] *adv.* 气动地

Exercises

Ⅰ. Answer the following questions according to the passage.

1. Can you tell me what the survey include?

   _____

2. What's the examination in drydock?

   _____

Ⅱ. Practic these new words.

1. English to Chinese.

   overhaul _____   surveyor _____

   stern frame _____   hatchway _____

   distortion _____   attachment _____

2. Chinese to English.

   隔离舱 _____   易腐蚀空间 _____

   制冷机械 _____   外壳镀层 _____

   阴极保护设备 _____   尾架 _____

Ⅲ. Explain the nouns.

1. berthing

   _____

2. surveyor

Ⅳ. Fill in the blanks with the proper words or expressions given below.

> together with, proved to, requirements, cofferdams, In addition, vapour

Liquefied gas tankers have _____ for annual surveys, as mentioned earlier and several additional items. All tanks, _____, pipes, etc., must be gas freed before survey. Where the maximum _____ pressure in the tanks is 0.7 bar or less the inner tank surfaces are to be examined. _____, the tanks must be water tested by a head of 2.45 m above the top of the tank. All tank level devices, gas detectors, inertia arrangements, etc., must be proved to be operating satisfactorily. The special survey requirements are as previously stated, _____ the examination internally and externally where possible of all tank areas. Tank mountings, supports, pipe connections and deck sealing arrangements must also be checked. Samples of insulation, where fitted, must be removed and the plating beneath examined. Pressure-relief and vacuum valves must be _____ be efficient. Refrigeration machinery must be examined.

Ⅴ. Translation.

1. Translate the following sentences into Chinese.

（1）The particularly severe operating conditions for an almost all-steel structure necessitate constant attention to the steelwork.

（2）During an annual survey the various closing appliances on all hatchways and other hull opening through which water might enter must be checked to be in efficient.

（3）The fire protection, detection and extinguishing arrangements for passenger ships are examined every year and for cargo ships every two years.

（4）For a special survey, the requirements of the annual survey must be met together with examinations.

(5) For tankers, additional special survey requirements include the inspection of all cargo tanks and cofferdam spaces.

2. Translate the short passage.

All ships must have an annual survey, which is carried out by a surveyor employed by the classification society. This survey should preferably take place in a drydock (Fig.14.1), but the period between in dock surveys may be extended up to 2.5 years, such an extension is permitted where the ship is coated with a high resistance paint and an approved automatic impressed current cathodic protection system is fitted, in-water surveys are permitted for ships which are less than 10 years old and greater than 38 m in breadth and have the paint and cathodic protection systems already referred to. Special surveys of more rigorous nature are required every 4 years. Continuous surveys are permitted where all the various hull compartments are examined in rotation over a period of 5 years between consecutive examinations.

Lesson 15　Quality Management

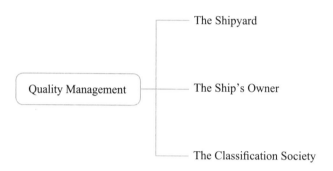

Background

A: Do you know what quality management is involved in the shipbuilding process?
B: It includes the shipyard, the ship's owner, the classification society and so on. Next we will study in further detail.

Text reading

　　The competition in the world is becoming fiercer and fiercer than ever before, especially in the shipbuilding industry. As for each enterprise, the quality of products has become her life line without any exception, and that is also the case with a shipyard. What is more, the quality of a vessel greatly concerns the safety of passengers and crew onboard. So any shipyard must not neglect the life-or-death matter of quality if she wishes to survive.

　　To begin with, we shall make clear several terms. They are management, quality assurance and quality control, with their short forms QM, QA and QC respectively. Among them, QM is a big idea while QC is a rather small, concrete and old one. In order to avoid any unnecessary misunderstanding, in this text, we will still keep to the old term QC. With the regard of QC, we shall first deal with the term TQC, which stands for the total quality control. The first implication of TQC is that the quality control covers the whole procedure of production from start to finish, even extending to the after service or the technical service. For instance, the moment a new order is accepted, TQC will go into action immediately; for, when the contract becomes validated and both parties have worked out the technical specification, quality has been involved already. The second implication of TQC is that it concerns all the personnel, including blue collar workers, technicians of workshops, staff members of such departments as design, supply , education and training and so on.

H.D. Shipbuilding Group buts extra emphasis on the quality management all the time. Her quality policy says, "Open up the market by first-rate products and win customers through fine quality and nice after-service". So far, the three main production lines in the group have been respectively granted the certificates of quality approval in accordance with ISO 9001 by the relative authorities. To put it bluntly, the ISO 9001 system itself will not improve the product quality but it will create a quality management procedure whereby the product quality can be monitored at all the stages so that the possible quality failure can be identified and traced back to Its source for the rectification of poor quality as well as for learning a necessary lesson for the future (Fig.15.1) .

Fig.15.1　H.D. Shipbuilding Group

If a vessel for export is to be built,things will get still more complex than to build a vessel for home use,since different parties are to be involved in the supervision activities.The parties concerned as follows:

1. The Shipyard

Under a shipyard, there will normally be the design department composed of marine engineers and naval architects, the quality control department whose representatives are called inspectors, the production department commanding various workshops whose representatives referred to as site architects will be sent to serve on the working site, the supply department, the financial department, ect. Except the financial department, all the above department are directly concerned with the shipbuilding quality.

For example, the supply department is in charge of purchasing the necessary material equipment, and requires makers to send out their service engineers to help install their equipment sold in accordance with the relevant contracts. As for the design department, it must be responsible for the basic design and the detail design as well as the completion drawings and the concerned information. In addition, it has to be always ready to answer the technical problems put forward on the spot of operation.

For the production department, however, stress is almost always laid on the working schedule and it keeps talking about CPM (the critical path method) and palletizing management all the time, whereas the quality control department will have a close look at everything installed onboard with a critical eye. It's difficult for inspectors to nod in approval. Therefore, a serious quarrel will break out between a site architect and an inspector from time to time.

2. The Ship's Owner

Any vessel is to be registered in a certain country and to be constructed, machinery istalled and equipment provided or finished according to the latest rules and regulations of a classification society. Besides ,the vessel is to comply with all the concerned international regulations , such as IMCO 1966 and SOLAS 1974.

Only when a vessel conforms to the above mentioned rules and regulations, can a shipyard obtain the relevant certificates from the classification society and international authorities.

In order to execute the supervision of shipbuilding, the owner or his representatives sometimes called as site supervisions will station in the yard. And , do remember that the owner, the contract in hand, is the king of the yard , and will always have the last word.

3. The Classification Society

There are a few famous classification societies in the world, such as NK, LR, ABS, DNV ,GL and CCS (Fig.15.2).

Fig. 15.2 CCS

The classification society involved will send out to the yard its registers for diesel engine making and its surveyors for shipbuilding, and they are to work on the basis of the contract, the specification and the concerned standards.

In short , the classification society would, if necessary, arbitrate between the owner and the yard, should a dispute arise.

The shipbuilding supervision covers materials equipment, drawings, construction and tests and trials including the shop trial, dock trail and sea trial.

With a successful sea trial, the vessel is at last nearly to be delivered. And, of course, the delivery ceremony is really a great occasion to all the parties concerned and after a long journey full of fierce quarrel, warm argument, mutual understanding and nice cooperation they will say "cheers" to each other and enjoy champagne together.

New Words and Expressions

1. fierce [fiəs] *adj.* 强烈的；凶猛的；酷烈的
2. enterprise [ˈeʌtəpraɪz] *n.* 企业；事业
3. neglect [nɪˈglekt] *vt.* 疏忽，忽视；忽略
4. assurance [əˈʃʊərəns] *n.* 保证，担保
5. validate ['vælɪdeɪt] *vt.* 使合法化，使有法律效力；使生效
6. blue collar worker 蓝领工人
7. H.D. Shipbuilding Group 沪东造船集团
8. emphasis [ˈemfəsɪs] *n.* 重点；强调
9. certificate [səˈtɪfɪkət] *n.* 证书；执照
10. supervision [ˌsuːpəˈvɪʒ(ə)n] *n.* 监督；管理；
11. marine engineers 轮机工程师
12. naval architects 造船工程师
13. purchasing [ˈpɜːtʃəsɪŋ] *v.* 购买 (purchase 的现在分词); 购买东西
14. Palletizing management 托盘管理
15. conform [kənˈfɔːrm] *vi.* 符合；遵照
16. the classification society 船级社
17. register [ˈredʒɪstə(r)] *n.* 登记，注册
18. diesel engine making 柴油机制造
19. arbitrate [ˈɑːbɪtreɪt] *v.* 仲裁
20. the shop trial 车间试验

Notes

1. As for each enterprise, the quality of products has become her life line without any exception, and that is also the case with a shipyard. What is more, the quality of a vessel greatly concerns the safety of passengers and crew onboard.

   译：对于每一个企业来说，产品质量无一例外地成为她的生命线，对于一个船厂来说也是如此。更重要的是，船舶的质量关系到船上乘客和船员的安全。

2. To begin with, we shall make clear several terms. They are management, quality assurance and quality control, with their short forms QM, QA and QC respectively.

   译：首先，我们将阐明几个术语。它们是质量管理、质量保证和质量控制，分别简称为 QM、QA 和 QC。

3. For example, the supply department is in charge of purchasing the necessary material equipment, and requires makers to send out their service engineers to help install their equipment sold in accordance with the relevant contracts.

译：例如，供应部门负责购买必要的材料设备，并要求制造商派遣服务工程师按照相关合同帮助安装他们出售的设备。

Expanding reading

Palletizing Management

Palletizing most often refers to the act of placing products on a pallet for shipment or storage in logistics supply chains. Ideally, products are stacked in a pattern that maximizes the amount of product in the load by weight and volume while being stable enough to prevent products from shifting, toppling, or crushing each other.

Palletizing management as one of the management methods of modern shipbuilding mode, is produced with the development of the manufacturing and installation methods of ship outfits. Palletizing management is a new management method introduced from abroad with the development of Shipbuilding Design in 1980s. As early as the 1970s, Japanese shipbuilding enterprises have already realized palletizing management in the design and management of outfitting operations. In Japan, palletizing management is the core of production management. Together with shipbuilding production design, reasonable production system and type of work structure and perfect planning work, palletizing management constitutes the four powerful pillars of internal production management of enterprises, which enables the rapid and efficient development of shipbuilding production in Japan. Palletizing management arises with the development of methods for manufacturing and installing ship outfits.

Fig. 15.3　Palletizing Management

In the process of ship production, palletizing is not only an operation unit, but also an assembly unit of parts, components and equipment for installation. There are a large number of outfits to be made and installed during the process of ship construction. In order to facilitate the construction and management, they are divided into smaller units according to the operation stage and workplace. Each palletizing, that is, a small project, is completed at the same stage and in the same place, and the commencement and completion dates of each palletizing project are clearly specified. The whole project plan, project schedule and supplies of materials and equipment can also be implemented one by one（Fig.15.3）.

New Words and Expressions

1. Palletizing Management 托盘管理
2. logistics supply chains 物流供应链
3. shifting [ˈʃɪftɪŋ] n. 移位
4. toppling [ˈtɔːp(ə)l] n. 倾倒
5. crushing [ˈkrʌʃɪŋ] n. 破碎
6. outfitting operation 舾装作业
7. project plan 工程计划
8. project schedule 工程进度
9. supplies of materials and equipment 物资器材供应
10. implement [ˈɪmplɪmənt] v. 执行，实施

Exercises

Ⅰ. Answer the following questions according to the passage.

1. Can you tell me what is the meaning of the quality management?
   _____

2. What role does the classification society play?
   _____

Ⅱ. Practic these new words.

1. English to Chinese.

   shop trial _____   dock trail _____

   sea trial _____   quality assurance _____

   quality control _____   quality management _____

2. Chinese to English.

蓝领工人 _____   托盘管理 _____

船级社 _____   柴油机制造 _____

沪东造船集团 _____   造船工程师 _____

Ⅲ. Explain the nouns.

1. marine engineers

_____

2. palletizing management

_____

Ⅳ. Fill in the blanks with the proper words or expressions given below.

> for instance, involved, stands for, immediately,
> quality control, to begin with, in order to

_____, we shall make clear several terms. They are management, quality assurance and _____, with their short forms QM, QA and QC respectively. Among them, QM is a big idea while QC is a rather small, concrete and old one. _____ avoid any unnecessary misunderstanding, in this text, we will still keep to the old term QC. With the regard of QC, we shall first deal with the term TQC, which _____ the total quality control. The first implication of TQC is that the quality control covers the whole procedure of production from start to finish, even extending to the after service or the technical service. _____, the moment a new order is accepted, TQC will go into action _____; for, when the contract becomes validated and both parties have worked out the technical specification, quality has been _____ already.

Ⅴ. Translation.

1. Translate the following sentences into Chinese.

（1）H.D. Shipbuilding Group buts extra emphasis on the quality management all the time. Her quality policy says, "Open up the market by first-rate products and win customers through fine quality and nice after-service".

_____

(2) As for the design department, it must be responsible for the basic design and the detail design as well as the completion drawings and the concerned information.

(3) Any vessel is to be registered in a certain country and to be constructed, machinery istalled and equipment provided or finished according to the latest rules and regulations of a classification society.

(4) Only when a vessel conforms to the above mentioned rules and regulations, can a shipyard obtain the relevant certificates from the classification society and international authorities.

(5) In short, the classification society would, if necessary, arbitrate between the owner and the yard, should a dispute arise.

2. Translate the short passage.

Among them, QM is a big idea while QC is a rather small, concrete and old one. In order to avoid any unnecessary misunderstanding, in this text, we will still keep to the old term QC. With the regard of QC, we shall first deal with the term TQC, which stands for the total quality control. The first implication of TQC is that the quality control covers the whole procedure of production from start to finish, even extending to the after service or the technical service. For instance, the moment a new order is accepted, TQC will go into action immediately; for, when the contract becomes validated and both parties have worked out the technical specification, quality has been involved already. The second implication of TQC is that it concerns all the personnel, including blue collar workers, technicians of workshops, staff members of such departments as design, supply, education and training and so on.